Education Division 2021

Held at the 2021 AIChE Annual Meeting

Boston, Massachusetts, USA and Online
7 - 11 November and 15 - 19 November 2021

ISBN: 978-1-7138-5700-6

Education Division 2021

Held at the 2021 AIChE Annual Meeting

Boston, Massachusetts, USA and Online
7 - 11 November and 15 - 19 November 2021

AIChE
The Global Home of Chemical Engineers

Printed from e-media with permission by:

Curran Associates, Inc.
57 Morehouse Lane
Red Hook, NY 12571

CURRAN ASSOCIATES INC. proceedings.com

Some format issues inherent in the e-media version may also appear in this print version.

Printed with permission by Curran Associates, Inc. (2022)

For permission requests, please contact AIChE
at the address below.

AIChE
120 Wall Street, FL 23
New York, NY 10005-4020

Phone: (800) 242-4363
Fax: (203) 775-5177

www.aiche.org

Additional copies of this publication are available from:

Curran Associates, Inc.
57 Morehouse Lane
Red Hook, NY 12571 USA
Phone: 845-758-0400
Fax: 845-758-2633
Email: curran@proceedings.com
Web: www.proceedings.com

TABLE OF CONTENTS

631290 Evaluating Performance in a Summer Online Thermodynamics Course Using a Concept Inventory

Abstract:

In our current context it seems essential to understand whether fully online course offerings are effective in achieving acceptable learning gains. Additionally, when teaching a foundational course over an abbreviated timeframe, instructors would benefit from confidence that learning gains are at least comparable to those from non-abbreviated offerings. This thermodynamics course is the sole course in its curriculum on the subject and covers 1st and 2nd law, equations of state, phase equilibrium, mixtures, and reaction equilibrium. It has a material and energy balances course as a pre-requisite. This presentation will summarize the offering and prior efforts at evaluation but will focus on the extension of evaluation of two cohorts by use of a targeted subset of a concept inventory (CI, by Vigeant et al.) for specific frequently misunderstood conceptual contrasts. Students were asked to complete the CI at the start of the course and then again and gains from pre- to post- were compared with previously published results using the same CI. Resulting gains in the summer course were comparable to prior results from full-term courses but lack statistical significance due to small numbers in the summer course offerings.

627784 Engaging an Entrepreneurial Mindset Though Open-Ended Projects in the Sophomore Year

Abstract:

An ability to view a task with an entrepreneurial mindset (EM) is a skill that can set engineering students apart. Having an EM requires students to demonstrate curiosity to find connections and create value when solving a problem. In order to develop EM, open-ended projects can be utilized where students have the freedom to think critically and creatively to solve a problem. In sophomore level Materials in Engineering Systems and Fluid-Thermal Systems courses, two projects were created to engage students in EM thinking while reinforcing course concepts using open-ended projects. First, in a Fall semester Materials in Engineering Systems course, students completed a project to choose a new material to redesign the molasses storage tank that failed in Boston in 1919. In the project, teams of students needed to consider the properties of different materials, the thickness of the storage tank needed based on their materials choice, and the material cost of this new tank. Then, in a Spring semester Fluid-Thermal Systems course, student teams were given different fluid flow system scenarios that they needed to design to determine a pipe material, pipe diameter, necessary fittings, minimum pumping power, and system cost. Minimal system constraints were provided to require students to explore alternative solutions, seek information from outside sources to acquire knowledge, and persist through potential design failure. Overall, students had a positive reception to these projects, and many expressed enjoying having the liberty to analyze materials and fluids that are not widely discussed in class. Project assignment sheets and implementation details can be found on cards published on the Engineering Unleashed Platform.

624079 Problem Solving When Textbook Problems Are Replaced with Student-Written Youtube Problems

Abstract:

Employers are seeking creative problem-solvers, while faculty are often challenged getting students ready for the industry. One way to improve problem solving ability is by practicing with homework problems. Text-based questions with industrial relevance may be engaging to students. However, using the same 'textbook' problems can lead to the solution manual dilemma where solutions are readily available online. Here, engaging students with interesting, relevant, and real world problems tie activities in a video to a homework-style problem, which are called YouTube problems. Leveraging established learning theories, visuals play vital role in creating long and short-term memories. Moreover, incorporating visuals in teaching have led to positive learning outcomes in engineering education. A YouTube problem is created when a student selects an online video, writes course-related problems based on the video, and creates a solution. Student-written YouTube problems can be applied as future homework practice or assessment. YouTube problems have been written for variety of courses over 5+ years, and we have an archive of more than 500 YouTube problems. The research here explored replacing textbook homework problems with student-written YouTube problems. To answer a series of research questions over two treatment and two control cohorts, we used three measures: 1. Problem solving using an established rubric PROCESS, 2. Perception of difficulty with NASA TLX survey, and 3. Learning attitudes through another survey known as CLASS. Overall, students have reported comparable rigor levels for YouTube and problems from textbook and displayed higher problem-solving abilities with YouTube problems. Students' attitudes increased toward engineering.

629597 Addressing Student Equity in the First Year through the Student Experience Project

Abstract:

The Student Experience Project (SEP) seeks to develop equitable learning environments for students on campus. The SEP launched its programs in fall 2019 with six public urban universities. The University of Toledo joined the program and enlisted instructors of first year classes to participate in fall 2020.

Our chemical engineering program is a mandatory co-op program. Approximately one-half of a student cohort leaves in the spring of the sophomore year for one of their three required co-op experiences. To prepare students for their first co-op, mass and energy balances are taught in the first year (mass balances in the fall and a continuation in the spring) and thermodynamics in the fall of the second. Moving these classes so early in the curriculum creates unique challenges as students make the transition to a university learning environment and bring a broad spectrum of mathematical and scientific abilities to the classroom.

I have been the mass balances instructor for several years and elected to participate in the SEP program to help improve classroom equity, especially given the impact of the ongoing COVID pandemic and racial strife. Given the diverse backgrounds of students, this is especially challenging for courses in the first semester.

The SEP program monitored student experiences through a set of five surveys during the semester. The surveys report student mindset in six areas:

1. Trust and fairness
2. Institutional growth mindset
3. Identify safety
4. Self-efficacy
5. Social belonging
6. Social connectedness

Results are reported for all students and six subareas:

1. High financial stress
2. Low financial stress
3. Structurally disadvantaged race
4. Structurally advantaged race
5. Male
6. Female

Building upon the extensive literature, SEP provides a toolbox for addressing student experiences in each area and incorporating activities that can enhance equity.

Experiences with the SEP program are reported. The tools used in hopes of improving equity are discussed. Both anecdotal and quantitative findings are presented and compared to other first semester classes on campus. The potential to integrate activities across multiple first year classes to improve student equity will be discussed.

625791 Teaching Process Calculations to Undergraduate Chemical Engineers: An Engaged & Energized Online Class

Abstract:

Introduction

"Process Calculations" is a required class in undergraduate ChE curricula worldwide. It serves not only as a good introduction to the subject but is also a pre-requisite for many ChE courses including Reaction Engineering, Separation Processes and so on. At IIT Madras, the course is in the first year of the four year undergraduate program, and sees a typical class size of ~100 students. Managing the stress levels of the students, alongside the learning goals for the course and the evaluations as required by administration, is a major challenge. In this talk, we want to illustrate a few best practices, applicable to online/hybrid classes in particular, with examples of student performance and testimonials.

Philosophy & guiding principles

In putting together the details of the course, we wanted to ensure that the first year undergraduate students felt a lot of excitement for chemical engineering. Second, examinations and evaluations were causing unprecedented levels of stress and competitiveness, and we wanted to do better with those, with the constraints of the existing systems. Finally, as the class was completely online, and the usual campus environment was absent, we wanted to shoulder an additional responsibility of building a sense of belonging to IIT Madras.

We are proponents of what we call "Play to learn" or "game based learning" at IIT Madras. Over the past few years, both in physical classrooms and in online ones, it has been our attempt to bring in more play and fun. In an elective course titled "Let us play to learn" we have students custom-building games and interactivities – these include board, mobile hypercasual, quiz-based & card games – for subsequent use in various classes. The ideas are gaining good popularity in both the student and faculty communities. We wanted to harness our learnings from this work, into a full-fledged core engineering course, and *Process Calculations*, as an introductory ChE core course, provided a great platform for the same.

Overall the goal of this teaching experiment was to provide an engaged and energized learning of the objectives of the course, to a class of 100+ students, in an online mode, using the broad guiding principles of game based learning.

Tools and Techniques

A detailed description of the evolution of the course is provided in this section. While several elements are common in chemical engineering pedagogy, the unique challenges due to completely online instruction, for a population that is not used to it, are addressed with new measures that are easy for instructors and learners to adopt. A few of the unsolved issues are highlighted as well, to enable further discussions, and future improvements.

- **Resource Materials**

The standard textbook used at IIT Madras for this course is the popular 'Elementary principles of chemical processes' by Felder and Rousseau. We found that it was particularly important to stick to the data tables, units, and graphical information in this book, to avoid confusions. In addition, a handful of short videos (10-15mins) long to described via animations, for the more difficult/confusing concepts were provided. All the lecture slides, and video recordings of the classroom interactions were made available to students as well.

- **Interactive Learning**

A common challenge faced by instructors in such large classes is getting students to interact – to ask or answer questions. The online mode further exacerbated this issue. It has been our finding that mixing in "fun" and "serious" questions, particularly via anonymous, online polling, is very helpful. Rating, word cloud, and open text modes of polls were also frequently used in class. A text chat-based discussion forum, worked very well in a live class, as well.

- **Bite Sized Fun Learning & Assessment**

We used short quizzes to provide well-rounding learning, in two ways. One, was a self-assessment tool, built like an online game with some fun graphics (bowling alleys to get full strikes at; mountains to climb, etc). The second was a graded "summary" quiz at the end of every class. Students routinely appreciated these interventions that helped them keep up-to-date on the course material, while having fun in a fail-safe manner.

- **Enabling Peer Learning**

A 30 mins breakout room session, during class, to enable students to discuss the tutorial solution methodology in groups, was provided. Students later individually solved the problems and uploaded their answers for checking by the teaching assistants. Possibly due to the students being in their homes (not on campus), the effectiveness of breakout rooms was not as good as could be.

- **Quick Feedback and In-depth Reflections**

Through the course of the semester, it was critical to not only remain open to receiving feedback from students, but also to act upon it with immediacy. The inputs regarding how the learning and progress were proceeding came up in various ways – a quick and immediate "reaction" (thumbs up/down, smiley/frowny faces) in the online class helped gauge if majority had understand the explanation provided. At the end of every week of classes, students were encouraged to answer open ended questions about the taught material, comment on their own learning and progress through the course, via an online form.

- **Assessments**

Three exams were conducted in the course. One was a rapid fire online quiz (with a leaderboard and some memes for fun), the second was based on typing numerical answers in an online form, and a final exam involving hand written solutions and uploading, were conducted. These exams were graded in the conventional manner, with emphasis on getting the right answer and the correct sequence of steps; and contributed 50% of the final grade, whereas the daily quizzes, class participation, reflections & tutorials contributed the rest. Overall, the goal was to give a judicious mix of points for effort and accuracy, while fitting into the constraints set by the institute administration.

Analysis

A few important motivating factors drive us to recommend these as good – if not best - practices to others. First, the level of interaction – in a typical in-person class, 10-20 students at the maximum answer or ask questions. The roster and specific calling out of students used here, quickly moved this number up to ~95 students (out of 118 total registered). Second, the reflections forms proved to be a great way to obtain in-depth feedback from students. Many of the comments could be used directly to improve the course experience immediately, in the next week.

Students frequently expressed their relief at being able to learn about and build on chemical engineering concepts, in a less judgmental, uncompetitive, fun manner. Balancing effort with quality, with tools used so students could self-evaluate (via games like the bowling alley one), were definitely ideas that bore rich dividends. The polls, were helpful to gauge the mood of the class, and served as good de-stressors as well.

Some of the tough parts of a Process Calculations course involve the use of diagrams – like the Psychrometric Chart in design of air-conditioning systems. While the use of the basic principles of mass and energy balances are well illustrated in such applications, they have often been a source of fear and frustration among undergraduate ChE students. With the game-based learning and very interactive mode of instruction used here, several students voluntarily stated that this was their "favourite" topic in the course!

Some of the final comments by students are shown below:

- *Well, I was not a very big fan of chemical reactions and inorganic chemistry. Earlier thought that chemical engineering=chemist. After this course, I realised that chemical engineering is just the perfect branch for me..*
- *This course taught a lot of practical things, and all these concepts and problem solving skills will really be very helpful to us as chemical engineers. The course structure was organized and that made it interesting to study it.*
- *The best part was interactive session and time to time quiz for regular evaluation.*
- *CH is the only course that doesn't feel disjoint from reality*

Some of the elements of the course that worked less well were – peer interaction and tutorials. In the initial part of the semester, the students were reluctant to speak in their breakout rooms. Scanning and uploading hand-written answers was also a challenge for those with poor internet connectivity/older devices. Light weight apps or online modules integrated into learning management systems are probably better in terms of uniformity of learner experience.

Conclusions

The Covid pandemic has led to a major upheaval in our teaching-learning. Forging friendships, co-operative learning, and development of professor-student bonds have proved particularly difficult in these times. Using the principles of game-based learning in the first ChE course of their curriculum, can be critical in retaining their interest and enthusiasm for the subject. The evaluations also indicate an upswing in the class average marks, underscoring the possibility of integrating the prescribed grading methodologies, within these new ideas. The various online tools opened the doors to several new means for energizing and engaging students, and it is our hope that these will find a regular place in the future.

627906 From MOOCs to Residential Instruction — Departmental Perspective on Digital Learning @ MIT ChE

Abstract:

The Department of Chemical Engineering at MIT has seen a rapid growth in digital learning activities over the past few years. In this talk, we will showcase completed and ongoing digital learning projects and discuss the department's strategic vision. We will present three Massive Open Online Courses (MOOCs) on edX: "10.50.1x Analysis of Transport Phenomena I: Mathematical Methods", "10.50.2x Analysis of Transport Phenomena II: Applications", and "10.S95x Physics of COVID-19 Transmission"; one Small Private Online Course (SPOC): "10.MBCx Math Boot Camp for Engineers"; and other residential projects using digital tools guided by research-based teaching practices. We will share the lessons learned from producing digital educational contents, including videos, graphics, and assessments. Finally, we will explain our departmental strategy for continuing our digital learning work both residentially and globally.

629036 Virtual Implementation of a Hands-on Learning Tool and Its Effect on Student Comprehension and Motivational Gains

Abstract:

Our project involves the national dissemination of highly visual hands-on learning tools focused on fluid mechanics and heat transfer principles to 44 institutions and branch campuses within the United States. Like many other educators, our team had to adapt the implementation protocols to accommodate remote learning during the COVID-19 pandemic. Rather than students working in groups with our hands-on learning tools, we created follow-along video implementations and supplementary tutorial videos. The videos allow students to complete the complementary worksheets associated with each hands-on learning tool while watching a graduate student explain basic concepts and collect real-time data with the hands-on learning tools. The supplementary tutorial videos are focused on an in-depth discussion of a single conceptual aspect of the learning tool. Across three remote-learning semesters, a total of 36 virtual implementations at 12 institutions were completed with approximately 630 chemical and mechanical engineering students. An asynchronous implementation method was used for 70% of the virtual implementations, while other instructors presented virtual material in a synchronous virtual setting with several allowing live, small-group discussion. At the conference, we will present conceptual and motivational assessment results to compare the effectiveness of virtual implementations versus traditional interactive hands-on implementations.

623434 Developing Simulation-Based, Student-Centred Educational Software in Process-Control Education and Training

Abstract:

Laboratory experiments demonstrating important underlying principles of process control are often challenging in their capacity for scalability towards larger-numbers sessions for students, and are often very time- and resource-consuming. Indeed, the recent rush to embrace online teaching and learning in view of the COVID-19 pandemic has accelerated very substantially since Spring 2020 an already-emerging theme of software-based exploration of key concepts in process control for students by means of virtual experiments.

The present contribution has as its central hypothesis and goal the ascertaining of how the complete implementation of virtual-laboratory environments (VLE) can effect positive learning outcomes of a student via constructivist pedagogical approaches. In so doing, a modularised MATLAB-and-Simulink VLE platform has been developed, accompanied by appropriate web-tools, which is focussed specifically on this constructivist approach to student-centred learning. To assess the effectiveness of this VLE, volunteers were split into two groups - the reference and VLE group, both of which took part in the same temperature-control exercise (via PID control). The reference group were exposed to the more traditional behaviouristic practical mode of learning, where they would pay close attention to the actual real-world experimental system and results at UCD Chemical Engineering laboratories; here, the material is gone through in a more prescribed manner, where the student has a certain level of engagement in their own learning, as per classical theory. In contrast, the VLE group encountered the equivalent specifically-designed virtual environment, where they worked in groups and individually to engage with a virtual version of the PID temperature-control experiment, based on the real-world system, where they can vary input parameters and controller settings and explore themselves the system response in time and Laplace domains, with trial-and-error approaches lending themselves to determining optimal settings.

Using equal time allocations to both approaches, it was set up - by design - that the only possible factor between the outcomes for both groups was the pedagogical approaches taken to training. Upon completion, quick MCQ examinations were undertaken to assess the groups' learning outcomes, and it was found that the VLE cohort displayed a generally superior level of understanding; the level of active self-engagement with the topic was found to correlate reasonably well with the degree of understanding. This suggests that VLEs, designed carefully and done well with appropriate software examples, can lead to positive and effective training experiences.

621678 Reinventing Unit Operations Laboratories Post-COVID-19

Abstract:

With the emergence of COVID-19 in Spring 2020, faculty were forced to quickly convert face-to-face courses to online offerings. While this transition was a challenge across all courses, the need to move a hands-on laboratory experience to fully online required significant course modifications. As a result, faculty were forced to focus on laboratory learning outcomes to ensure that student learning goals were met in this non-traditional setting. Surprisingly, this opportunity to reflect on learning outcomes allowed many of the weaknesses of "traditional" unit operations courses to be identified. We will present our assessment of chemical engineering laboratory learning outcomes and how they are currently being satisfied in a traditional face-to-face unit operations course. To better meet the laboratory learning outcomes moving forward, we will discuss strategies for modernizing the teamwork, communication and problem-solving skills being taught in the lab through reinventing the unit operations course curriculum. Additionally, we will discuss two potential surveys: one to probe a broader set of unit operations instructors, and another to probe industry professionals. These surveys will ensure that curriculum changes meet both the standards and needs of academics as well as the ever-changing needs of engineers in industry. With this work, we aim to modernize the unit operations curriculum to better prepare students for their careers as working engineers.

633062 Teaching How to Fish: Mentoring Students in Job/Internship Networking

Abstract:

The job and internship market for chemical engineers is constantly changing with the economy and due to forces much beyond the control of Junior and Senior Chemical Engineering students. Here we report our efforts over the past decade seeking to teach an effective way to mentor students in effective targeted networking strategies and throughout the job search. We found that creating multiple seminar classes assisting students through the process was the most effective approach. This has resulted in increased student placement rates, with close to 100% placement within a few months of graduation in most cases.

632378 Teaching Technical Writing to Chemical Engineers

Abstract:

The ability to write clear and precise technical reports to a broad audience is a fundamental skill that Chemical Engineers develop while completing their undergraduate degrees. In a recent AIChE Education Division Survey, assessment of both oral and written communication is common to nearly all Chemical Engineering programs. Yet despite this ubiquity, and the perceived value of technical writing as a skill, it is often not considered a conventional engineering skill and this can affect the quality of writing instruction and thus student performance. At the University of Virginia (a public research institute), we noticed an increase in the number of 3rd year students entering our Unit Operations Lab who were unprepared to write technical engineering reports. A review of both our curriculum and teaching methods revealed a problem with how technical writing was being taught. First, a gap in writing instruction existed between 1st and 3rd year, and second, our own writing pedagogy was not well-matched with current best practice, nor was it designed to bridge this gap in instruction. Our solution was to re-design how we taught technical writing as part of our Unit Operations course, and to do so using existing resources. In this talk, I will present the design of our educational intervention, discuss the impact that the intervention had on the quality of student writing, and comment on the broader implications of our work for teaching the development of engineering skills in general.

Methods for teaching technical writing vary between institutions and are often a function of time and resource. A variety of formats (e.g. video tutorials, dedicated writing instructors) and methods (e.g. peer evaluation, instructor feedback, provision of sample reports) have been described for teaching writing, yet much of the literature focuses on 1st year Chemistry and Physics lab courses. Building upon this literature, we designed an intervention that targeted 3rd year Chemical Engineering students enrolled in a Unit Operations Lab. The intervention used a step-wise scaffold approach to teach the basic elements of writing a full technical report. Students wrote a series of brief individual memorandum that introduced these basic elements guided by detailed grading rubrics and documentation describing each report element. We hypothesized that this step-wise intervention would improve the overall quality of student technical writing and to test this hypothesis we compared the writing performance between two groups of students — a control group (n=30) (pre-intervention) and an experimental group (n=30) (post-intervention). An established writing assessment tool was used to analyze writing samples, and our results showed a statistically significant improvement in writing quality for the experimental group compared to the control group, as well as a statistically significant improvement in writing quality over time within the experimental group (i.e. pre/post mid-term). Our intervention improved the quality of students' written assessments and did so without increasing the workload for student nor instructor alike. The use of memorandum (instead of full technical reports) decreased the overall grading burden on instructors and demonstrated that improvements to pedagogy do not necessarily need to be additive. This 'less is more' philosophy has significant implications for how engineering skills can be taught, and this presentation will also discuss how we are adapting skills-development instruction elsewhere in the department.

629824 A Spectrum of Beliefs: A Study of Undergraduate Learning Assistant Reflections As Novice Teachers

Abstract:

Teaching complex chemical engineering classes like thermodynamics, transport, or reaction kinetics effectively requires sophisticated instructional design as well as many in-the-moment decisions during instruction. Attention has been paid to instructional design issues often focusing on content, but more recently including pedagogical strategies such as active learning as well. Less examined are the instructional choices that are made during instruction. In this study, we posit that these in-the-moment choices during instruction are based on an instructor's beliefs about teaching. In this qualitative study, we examine, in detail, the beliefs that two cohorts of undergraduate Learning Assistants (LAs) express about their instructional practice. One cohort taught during in-person instruction and one during remote. While these beliefs are specific to the context and role of the individuals studied, we argue that to the degree that the expressed beliefs represent the culture of chemical engineering teaching practice, a close examination of the LAs' beliefs can inform instructors in other chemical engineering settings.

For the past six years, the School of Chemical, Biological, and Environmental Engineering (CBEE) at Oregon State University (OSU) has integrated Undergraduate Learning Assistants (LAs) to support delivery of the core undergraduate courses in the curriculum. The LA Program utilizes the three core elements suggested by the *Learning Assistant Alliance* (https://www.learningassistantalliance.org/). First, in the LA Pedagogy Seminar, LAs receive pedagogical development in in a formal class with their peers, generally in their first term as an LA. Second, LAs meet weekly with the instructor and the graduate teaching assistants as a member of the instructional team to prepare for active learning in class. Third, LAs facilitate active learning of small groups and individuals in the class to which they are assigned. While LAs elsewhere are often used in large lecture sections, in the context of the unit studied, the LAs facilitated small group collaborative learning in smaller studio or laboratory sessions.

As part of the LA Pedagogy Seminar, the students are assigned weekly readings and then asked to reflect on the reading in relation to their own practice. The study reported here examines student responses at the start and end of their first term as an LA. Weekly reflection were collected from twenty-four LAs during the pedagogy course during in-person instruction and eighteen LAs the following year during remote instruction due to the COVID-19 pandemic. All participants provided informed consent to be in the study. LAs responded to the prompt, "*Think about teaching and learning in the environment that you are working in as an LA (e.g. studio , lab, ...). Identify two or three strong beliefs that you have about your role to support learning in that environment.*"

We ask the following research questions:

1. What are the common elements of the beliefs that LAs bring to instruction? How do LAs differ in their expressions around these beliefs?
2. How have the expressed beliefs shifted over the first term of the LA experience? What concepts from the LA pedagogy course are incorporated into the second reflection?
3. Are there substantial differences between the cohort teaching in-person and the cohort teaching remotely?

Thematic coding was used to characterize the aspects that the LAs identify as salient to teaching, the ways their thinking has shifted, and what concepts from the LA pedagogy seminar are being incorporated into their thinking. The beliefs were consolidated into five major themes: Instructional Climate, Group Work, Feedback, Student Thinking, and Course Content. Students bring different histories-in-person to their instructional role and we discuss these different perspectives using a histories-in-person lens.

631303 Using Virtual Learning Communities for Remote Learning in Large Lecture-Based and Flipped Classrooms.

Abstract:

One of the biggest challenges students faced during the COVID-19 pandemic was a loss of both social and learning communities. We examined the question of whether a virtual community within a class could serve some of the same purposes as their in-person communities. Our hypothesis was that putting students into virtual learning communities would encourage them to attend and participate in class and feel more supported in their learning. To test this hypothesis, we used learning communities in 2 large (>100 students) chemical engineering courses during remote and hybrid learning. The fluid mechanics course was taught in a lecture- based format while the material balances course had a flipped format. Both courses had full remote and hybrid modalities throughout the semester. Students stayed in their learning communities for either half of the semester or the entire semester and were put into breakout rooms with their community for many activities including: 1. brainstorming ideas on a shared google doc 2. Researching a topic such as types of valves and their uses and reporting back to the class, 3. working on in-class prolems together as a group, 4. Taking a practice test together. Peer evaluation of the group members was worth 5% of the final grade and focused on contributing to the learning of the group members in a safe environment and anonymous feedback was given throughout the semester. This presentation will report on surveys given to the students. Data from these surveys show that the learning communities created a sense of community among the students in the class while also holding the students accountable for participating during class. It will also discuss extending these ideas beyond the pandemic.

620542 Design and Analysis of a Sars-Cov-2 Vaccine Facility in a Remote Learning Format

Abstract:

We present here a novel case study for senior ChE capstone design project, where students work in groups remotely with very limited virtual instruction to design a manufacturing facility for one of four SARS-CoV-2 vaccine products. Each group of 3-4 students presented via PowerPoint and MS Teams on their progress four times and lead three formal meetings (also via MS Teams). There were 4 different products assigned to groups (either Pfizer/BioNTech, Moderna, Oxford-AstraZeneca, or Johnson & Johnson) as well as a specification for the facility bioreactors to be SIP/CIP or fully disposable. Student work product aside from oral presentations included three memos (summaries of what they communicated to their "supervisor" in meetings) and a written report with technical descriptions of their proposed facility, with emphasis on the economics, environmental impact, and safety. Students produced their batch process Gantt charts and flowsheets (which utilized bioreactors fermenting E. coli, HEK393A, or PER.C6 cells) in SuperPro Designer or MS Excel.

628545 Products from Pandemic Labs: Custom at-Home & Benchtop Unit Operations Teaching Modules

Abstract:

Departments across the country successfully addressed the COVID-19 pandemic with a wide variety of creativity. Like many others, our department accommodated students who were necessarily off-campus, in quarantine due to illness, exposure, or being in a particularly vulnerable population. Our department's teaching laboratories, however, were hit particularly hard by the fact that we were transitioning to a junior year unit operations series through 2021, and our last cohort of seniors were still on the old catalog year. These juniors and seniors would have been able to use the existing facilities in a normal year, but, with double occupancy, the need for social distancing required us to accommodate an inordinate number of students in chemical engineering laboratory activities, outside of a laboratory space.

To address this need at minimal cost and at scale, we relied heavily on custom 3D-printed and laser-cut parts, along with low-cost sensors, displays, and microcontrollers. All material costs were kept well under $50 for the majority of modules, and the activities replicated much of the experimentation, physical measurements, and data analysis of our existing unit operations equipment. Of course, safety was also of particular concern in module development, and each module was kept "kitchen safe."

The process equipment developed in this work include pumps (centrifugal & peristaltic), packed beds, venturi and orifice meters, solar hydrolysis units, mason jar CSTRs, and more. This talk will describe the resulting teaching module collection, and how they have been used by students in a chemical engineering curriculum.

We are all hopeful that upcoming semesters will not require such tools or measures, and all students will have access to the irreplicable experience of operating pilot-scale unit operations equipment. However, the teaching modules developed in this work will find substantial use post-pandemic. Their potential will be discussed for K-12 outreach, traditional chemical engineering laboratories, in-class demonstrations, and in continued support for students who may need to be off-campus for extended periods.

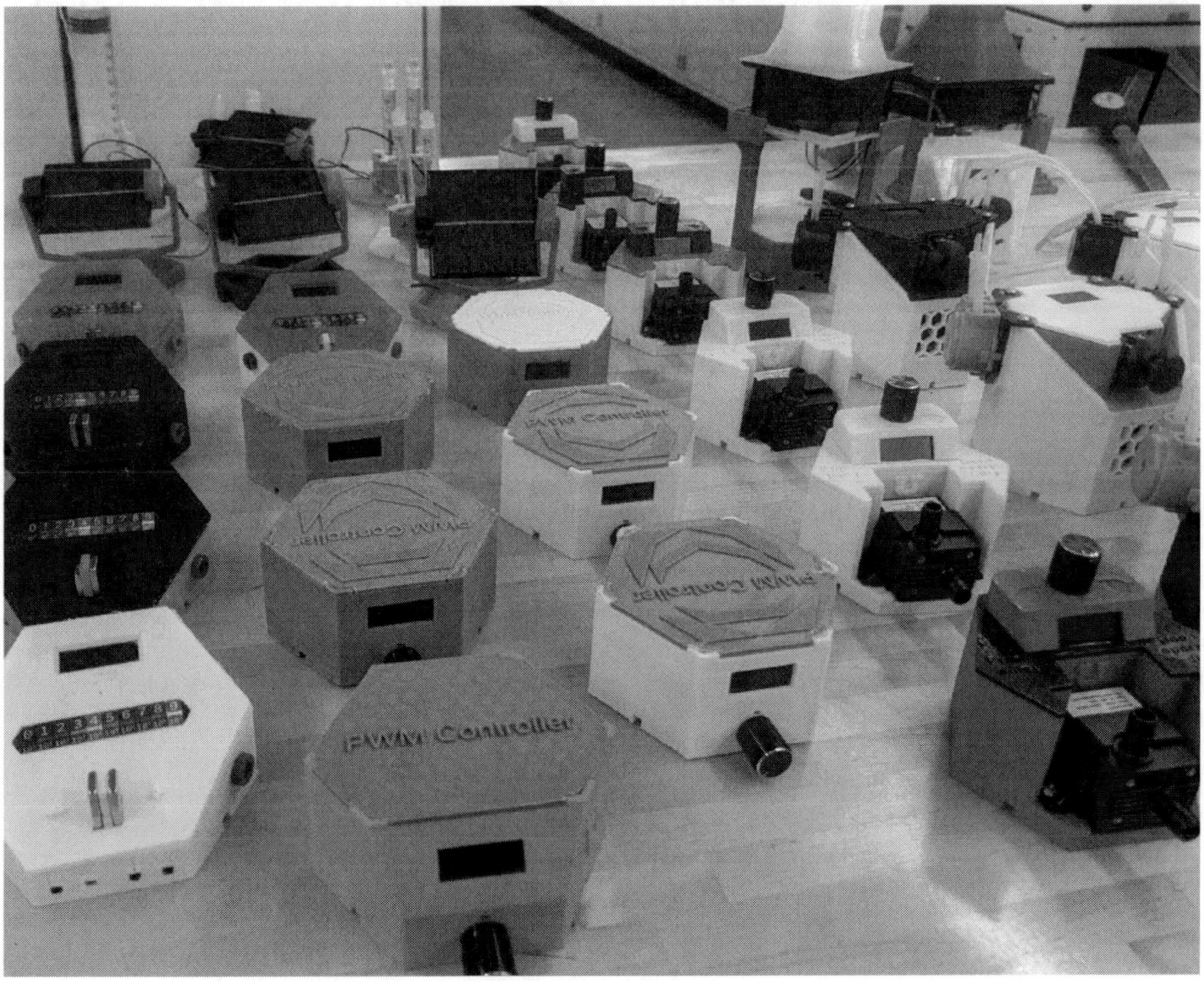
PWM Controller
PWM Controller
PWM Controller
PWM Controller

620015 Taking Cheme-Sports from on-Site to a Live Virtual Competition

Abstract:

ChemE-Sports^{TM} is part of the chemical engineering culture and has now gone virtual. Inspired by the success of the ChemE-Sports^{TM} in 2019 at the AIChE Annual Meeting in Orlando, an academic and industry team of a faculty member at Columbia University Chemical Engineering and Simulation Solutions Inc. developed a new method of using simulators for an on-line competition that can serve as foundation for chemical engineering competitions of the future. This international competition was open to universities as well as industry teams. The competition was created around the safety and optimization of a pentane-hexane distillation simulation owned by Simulation Solutions Inc. A competitive team bracket was developed for the competition during the virtual AIChE Student Conference on Sunday 15 November 2020. The results of the 2020 competition and lessons learned are presented here. This presentation will focus on the competition and modifications for remote participation from student and industry groups around the world.

637074 Accelerated Product Development and Commercialization in response to the Covid-19 Pandemic: Challenges and Opportunities

Abstract:

Biopharmaceutical industry is constantly looking to shorten product development cycle time and costs. The covid-19 pandemic provided unique set of conditions for the industry to push the boundaries in rapid product development and commercialization. In response to the pandemic, AstraZeneca is
developing a long-acting antibody combination being evaluated for prophylaxis/ treatment of Covid19. The program was highly accelerated to deliver batches manufactured at commercial scale in less than 6 months after program initiation. The presenter served as Operations technical lead, helped establish commercial manufacturing supply chains, and provided oversight of large-scale drug substance manufacturing for this program. Perspectives on supporting an agile and flexible team to deliver their best under pressure, lessons learnt, and opportunities to leverage learnings for accelerated biologics product development in the future will be presented.

630428 Mechanisms of Protein and DNA Removal in Depth Filtration

Abstract:

The routine use of depth filtration for harvest clarification in bioprocessing is aimed primarily at removal of particulates. However, impurities such as host-cell proteins and DNA are also typically removed to a significant extent. This presentation will explore some of the mechanisms that contribute to such impurity clearance. For proteins, we have measured the adsorption of model and therapeutic proteins on different depth filter components and assessed the driving force for such adsorption. We have also used confocal microscopy to determine the distribution of different proteins on the different depth filter components. Although filter aids typically behave as cation exchangers, protein adsorption is more commonly dominated by the binder, which has some anion-exchange character. For DNA, we have used covalent labeling of DNA and visualization by confocal microscopy of the extent of penetration of the retained DNA on depth filters. This has allowed resolution of retention by adsorption from retention by size-based filtration.

631392 Modeling Flux in Tangential Flow Filtration Using a Reverse Asymmetric Membrane for Chinese Hamster Ovary Cell Clarification

Abstract:

Tangential flow filtration is advantageous for bioreactor clarification as the permeate stream could be introduced directly to the subsequent product capture step. However, membrane fouling coupled with high product rejection has limited its use. Here, the performance of a reverse asymmetric hollow fiber membrane where the more open pore structure faces the feed stream and the barrier layer faces the permeate stream has been investigated. The open surface contains pores up to 40 μm in diameter while the tighter barrier layer has an average pore size of 0.4 μm. Filtration of Chinese hamster ovary cell feed streams has been investigated under conditions that could be expected in fed batch operations. The performance of the reverse asymmetric membrane is compared to that of symmetric hollow fiber membranes with nominal pore sizes of 0.2 and 0.65 μm. Laser scanning confocal microscopy was used to observe the locations of particle entrapment. The throughput of the reverse asymmetric membrane is significantly greater than the symmetric membranes. The membrane stabilizes an internal high permeability cake that acts like a depth filter. This stabilized cake can remove particulate matter that would foul the barrier layer if it faced the feed stream. An empirical model has been developed to describe the variation of flux and transmembrane pressure drop during filtration using reverse asymmetric membranes. Our results suggest that using a reverse asymmetric membrane could avoid severe flux decline associated with fouling of the barrier layer during bioreactor clarification.

631418 Filtration of Lipid Nanoparticles: The Importance of Understanding Mechanisms to Improve Performance

Abstract:

In bioprocessing, membrane filtration is used to remove microorganisms from the process stream and to assure the drug product is safe for administration to patients. Sterilizing filters are used to ensure that bacteria do not enter the process stream and that any bacteria present are removed. Most critical is the final 0.2 μm sterilizing-grade filter, which ensures sterility of the final product.

Oily particles, such as the lipids used to encapsule mRNA in some COVID-19 vaccines, have some unique characteristics that can create challenges for sterilizing filtration. Bacteria retention in streams containing oily particles is particularly sensitive to variables that do not impact retention in aqueous streams. Understanding the operating parameters that impact filtration throughput and microbial retention in streams containing oily particles will enable robust process design.

In this study, we review potential mechanisms for the challenges observed during sterile filtration of oily particles and consider the impact of membrane, bacterial challenge, feed composition and process conditions. Data will be presented on factors that can impact capacity, retention, or both. Perhaps the most significant factor is a pressure-dependent, pore blockage phenomenon unique to oily particles. During process design, properties such as processing pressure and temperature, particle size, and membrane selection should be considered to mitigate the challenges that can arise during processing.

By understanding the process and the unique properties of oily nanoparticles, robust processes can be designed for therapeutics or vaccines that rely on these particles as adjuvants and delivery vehicles.

631200 Replacing Exams with Projects: Advantages and Disadvantages As Observed during Hybrid and Remote Learning and Broader Efforts Towards Equitable Assessment

Abstract:

In response to a transition to remote and hybrid learning from the COVID-19 pandemic, as well as an environment with generally higher levels of stress that can cause students difficulty to prove their knowledge and ability, many instructors have tried to find ways of changing their means of evaluating student performance and understanding. The instructors' efforts have also been impacted on searching for methods that are more equitable to students of all backgrounds and experiences, and so the assessment reflects less inherent bias. Some methods have included weekly quizzes as opposed to larger exams, oral discussion exams instead of written tests, and replacing exams with larger projects. The efforts being made to broaden potential approaches to evaluation may have short-term benefits to provide workable assessment tools for the current environment, but will also provide insight into the effectiveness of each approach to be considered when regular in-person instruction begins again.

At Northeastern University in both a conservation principles course and a fluid mechanics course, students were assigned a short project in addition to each exam, with each project due date being within a week of its respective exam. The project allowed students the opportunity to present their conceptual understanding and knowledge in a creative format, being able to make a comic, video, or simulation addressing concepts from that part of the course. The projects provided a means for students to either reiterate their grasp on the course concepts in either a follow-up or lead-in to their performance on the exam, or to better show their understanding in a different, less-stressful, more open format in potential response to their previous performance. The introduction of the projects also served to reduce the number of exams in the semester, and could be completed remotely.

In this work, we discuss the impacts of these efforts in both short and long semesters (7 week vs. 14 week schedules), student performance and feedback, and insight into the advantages and disadvantages of this approach.

630363 Lessons Learned during the COVID-19 Pandemic: A Literature Synthesis

Abstract:

The COVID-19 pandemic brought a widespread shift in instructional practice as faculty scrambled to shift to remote instruction. One positive consequence is that this shift has inspired many instructors to think carefully about their teaching practice and the ways that technology tools and instructional strategies support their learning goals. Correspondingly, STEM educators have been investigating these needed changes and reporting them. There have also been coordinated efforts within our community to support chemical engineering educators such as five AIChE EdDiv Cirtual Communities of Practice. A corresponding spike in discipline-based education research (DBER) has also occurred. For example, over 250 papers have appeared in three archival journals related to chemical engineering: *Advances in Engineering Education*, *Biomedical Engineering Education*, and the *Journal of Chemical Education* (*Chemical Engineering Education* will be also publish a special issue in January 2022). Each of these reports focuses on specific contexts, mostly at the course level. In this paper we report a synthesis of the literature to identify how educators responded to the pandemic, and what we can learn from it. Topics analyzed include: course design, content delivery, assessment practices, classroom environment, project and laboratory work, the student experience, and social justice.

630756 Interventions to Mitigate Loss of Community during COVID-19

Abstract:

The switch to remote learning in response to the COVID-19 pandemic caught everyone by surprise, but faculty worked together to quickly transition. The administration worked to help student set up home learning stations by loaning them equipment and providing IT assistance. In terms of mechanics, everything seemed to be working. Students quickly learned, however, that webcams were overrated. Live-streaming video destabilized their already weak internet connections. Coupled with Wi-fi frustration is the requirement that you and your roommates must be dressed and the room be some semblance of clean. Students rarely turned on their cameras. For faculty this meant talking to a bunch of black squares with names. For students, they lost their peers.

In Spring of 2020, the loss of community was overlooked. Students' established study groups and communication channels for the semester were challenged by the campus closure. Though functional, academic life had become strange and unsatisfying. The Fall 2020 classes were a different experience. Although there were still technology challenges, support systems were in place to provide lectures, host discussions and even to perform lab experiments safely. The university was most concerned about the social interactions amongst the freshmen and worked vigorously to build relationships among peers. Seniors generally had well-established relationships with their classmates. Sophomores still had their freshman networks but were beginning to branch off into their selected majors. The juniors were most affected by the loss of community. Their friends who were in other majors did not understand the challenges they were facing with transport phenomenon or reactor design. Many juniors dropped their most challenging courses or withdrew from school to wait until in-person classes were available. An unusually large number of students landed on academic probation.

As associate chair for undergraduate studies, I reached out to all students on academic probation and invited them to set up weekly Zoom meetings with me. Two accepted my invitation in Fall 2020. During our 5- to 15-minute sessions they talked about their academic and home lives without judgment from me. I was able to point out resources they may have been unaware of, but generally tried to avoid giving unsolicited advice. I encouraged them to find something to look forward to each week. By the end of the semester, they had both completed all courses successfully and were no longer in academic jeopardy. I repeated this for Spring 2021 and had three students sign on with similar success. In the presentation I will look at common characteristics of those who chose to meet with me and why this was helpful to them.

In Spring of 2021, the lack of community was most noticeable. In one junior-level course, the instructor and course content had not changed, but the number of students struggling with the course had increased dramatically. The number of students complaining about the workload or lack of support in all of our junior-level courses had increased. In conversation with these struggling students it seemed that they were suffering from an absence of people to hold them accountable. Nobody was asking those simple questions like "Where were you last Tuesday?" or "Have you started on the homework yet?". It was apparent that the juniors had not formed study groups within their major before the pandemic hit. In cooperation with the Omega Chi Epsilon chapter, the department established FLASH sessions (Facilitating Learning Activities with Student Help). Senior members of the honor society open a classroom for an in-person study hall. Occasionally the students who attend are looking for help with homework. More often, they just crave conversation. Although attendance has been small, these sessions have made an impact on the students that attend.

In this presentation, I will discuss these two types of intervention and how we have continued these into Fall 2021.

624914 Oral Exams in Undergraduate Cheme Thermo Taught Online

Abstract:

COVID-19 pandemic challenged educators in many ways, and in particular with a need for new means of students' assessment. Instead of trying to modify conventional written tests with advanced proctoring technologies, we proposed an alternative exam format, and tested it within the undergraduate Chemical Engineering Thermodynamics II course, taught synchronously online. The assessment was based on two parts: a "take-home" written part, somewhat similar to conventional written tests, and an oral defense of the written part. The oral part of the exam was carried out as individual WebEx meetings with students, and included a number of concept questions on the exam problems. The weight of the oral part in the overall exam grade was comparable to that of the written part. The effectiveness of this approach was demonstrated based on the consistency of the students' grades with other means of assessment, and comparison with the student performance in previous years. Additionally, the approach was evaluated based on several student surveys, which showed that many students liked this format, and found oral sessions helpful to feel more involved with the course. Furthermore, the vast majority of the students found this format of the exam helpful to learn the course material. Although this approach does not save the instructor's time spent on assessment, the students benefit from this time more. We conclude that our approach is a viable assessment option for classes below 20 students, or larger classes if teaching assistants/proctors are available. This approach can be used in the courses taught both online and in-person.

624774 Online Cheating: A Darker Side of Educational Technology

Abstract:

While technology has had many positive impacts on education, particularly during the COVID-19 pandemic, it has also offered new opportunities for academic integrity violations. In particular, "tutoring" websites such as Chegg and Course Hero can be used by students to post homework or even exam problems for another person to solve; these problems and solutions are then posted online, possibly violating copyright laws as well as codes of student conduct at the students' academic institutions.

This study examines the prevalence of custom instructor-created exam problems posted to tutoring websites without instructor knowledge. Findings will include the frequency of exam problems appearing online and whether the nature of the course (traditional, online, etc.) has any effect on frequency. Best practices toward (a) identifying whether students have posted materials online and (b) minimizing opportunities for students to post course materials online will be presented.

622899 Aiche Education Division Virtual Communities of Practice (VCPs): Updates from the 2020-2021 Academic Year

Abstract:

Since the campus closures associated with the COVID-19 pandemic began in March 2020, the AIChE Education Division has provided a supportive network for faculty development through the initiation and implementation of the Virtual Communities of Practice (VCP) Program. This program consisted of five communities focused on different areas of the chemical engineering curricula: (1) Material & Energy Balances / Thermodynamics, (2) Transport / Separations, (3) Reactors / Controls / Computing, (4) Design, and (5) Lab. Detailed descriptions of these communities, including survey analysis and results obtained from March to December 2020, have been previously published in *Chemical Engineering Education*.

The VCP program continued throughout the Spring 2021 semester and an additional survey was administered to participants to further study the impact of this program on different aspects of faculty professional development. Results aggregated from over more than one year of VCP meetings will be provided to highlight the impact of this program on the AIChE community. Plans for future iterations of the VCP program and associated faculty community-building activities will also be discussed.

629706 Bourbon Production Engineering: Teaching Cheme Concepts to Anyone through the Science of Bourbon

Abstract:

Over the past 20 years, bourbon production has quadrupled, with 95% of the world's bourbon supply being produced in the state of Kentucky. As a result, distilleries have undergone significant expansions resulting in increased workforce demands. Additionally, increased product demands have necessitated the improvement of process efficiencies, resulting in the hiring of an engineering focused workforce. In 2019, the University of Kentucky formed the James B. Beam Institute for Kentucky Spirits with the commitment to support the growth of the Kentucky spirits industry through innovation in research and development of the state's distillery workforce. With the development of this institute, a focus has been placed on curriculum development to train undergraduate students for work in the distillery industry as well as provide continuing education for existing workers. As a result, a Bourbon Production Engineering course was developed which focuses on the teaching of chemical engineering concepts to students and workers from outside the field of chemical engineering. This presentation will provide an overview of the bourbon production process, with an emphasis on how bourbon production can be used to teach chemical engineering concepts to a general audience. Key concepts integrated into the course include material balances, heat transfer, diffusion and separation processes. By the end of the course, students have the skills necessary to design and estimate the costs associated with a bourbon production facility, including modeling a distillation column through basic McCabe-Thiele analysis.

629693 Re-Envisioning a Faculty-Led, International, Chemical Engineering Experiential Program in a Virtual Format

Abstract:

Since March 2020 the COVID-19 pandemic has diminished all travel-abroad educational opportunities that universities can offer their students. This presentation discusses the development and execution of a virtual, six week intensive, faculty-led, international program with Brazilian partners focused in sustainable energy during the months of May and June 2020. The program introduced chemical engineering students to global sustainability issues and how to produce innovate solutions to these problems using chemical engineering concepts. During this six-week virtual program, the students took two courses, Conservation Principles and Sustainable Energy Systems. In addition to the courses, they participated in discussions, activities and a project that deepened their understanding of cultural and societal differences in the way Brazil approaches sustainable development. Three main program strategies were developed. First, there were virtual lectures from representatives working in Brazilian companies. The representatives discussed their sustainable energy technologies and how the companies make an impact on the environment, society, and economy both before and during the pandemic. Secondly, the students took part in weekly culture workshops and discussions to highlight Brazilian culture. The topics included history, music, cuisine, and language. The students performed research and discussed these aspects of Brazilian culture and how they influence sustainable practices and policies. Lastly and most importantly, students worked with Brazilian engineering students from Maua University to perform a project in collaboration with Suzano S.A., a sustainable pulp and paper company located in Brazil. Students were divided into 5 teams consisting of a mixture of Northeastern and Maua students. They performed research, developed, and designed a plan in which Suzano can become more sustainable in a designated sector of business of interest to Suzano. Each team work together to gain and apply technical engineering knowledge to their sector's assignment. The team also accessed the impact that their solution had on the sustainability of the company. The Northeastern students gained valuable global knowledge and skills via virtual means working with the Maua students and company leaders.

A student close out survey found that 50% of the students (N=10) said program exceeded their expectations and the other 50% said that it met their expectations. This paper discusses strategies on how to successfully perform an international learning experience virtually. It will also discuss the types of assignments and interactions that ensure students engage in cross cultural learning while applying engineering and sustainability topics in a virtual environment.

623546 A Sustainability and Alternative Energy Course As a Bridge between Disciplines

Abstract:

Knowledge of sustainability, climate change, and environmental impact is far more important now than when the chemical engineering curriculum was created and refined in the early 20th century. Making room for these essential topics is going to be an important part of how chemical engineering as a course of study (and as a profession) survives as we navigate our current global pollution epidemic.

Our work here aims to help fill this gap by providing students with an introduction to sustainability and sustainable development, then teaching them ways that we are attempting to reimagine our energy system in our future near-zero carbon world. The course then moves to the technique of life cycle assessment as an important (but approachable) concept for applying the laws of conservation of mass and energy to products and processes to estimate their relative merits. As the problems we face are interdisciplinary, we tried as much as possible for a diverse cohort of students from all schools at The Cooper Union (Architecture, Art, and Engineering) to inspire discussion and different ways of thinking around the important political, cultural, and technical issues that must be navigated. Course content and structure was aimed to be as interactive as possible, including extensive in-class discussion activities as well as two large group projects where students presented their findings to the class.

We hope to demonstrate that our students gained an increased understanding of both the science and technology in energy system choices and an appreciation for the ethical trade-offs. Our course was offered in an entirely remote format, offering additional challenges for a subject requiring discussion and project work.

629821 Even More Steal-Able: Thermodynamics Concept Laboratories & Projects

Abstract:

Chemical Engineering Thermodynamics is a one-semester four-credit course with weekly laboratories. The course addresses energy and entropy, modeling of fluid behavior, and reaction thermodynamics. Halfway through Spring 2020, the course abruptly moved online in emergency remote instruction. In the second Spring semester of pandemic-constrained teaching, some students were attending class remotely, either by choice or due to quarantine, and some attended in person. Because this was known in advance, it became possible to take a a more considered approach to remote and hybrid laboratories and in-class problem solving activities. This talk will demonstrate several labs and in-class problems that were redesigned to work similarly for both in-person and remote students.

In the course's conceptual experiments, every activity starts with a prediction and ends with a reflection. Preserving these from the hands-on version, we were able to use simulations for some experiments and videos for others. We will discuss the relative effectiveness of these modes, as demonstrated by pre-/post-semester concept inventory as well as student responses to laboratory questions. While the learning gains for hands-on experiments appear larger, it is likely the results are also influenced by the same factors led to the student participating remotely that day. Now that every lab and class meeting has simulations or videos that set-up the experiment or problem of the day, I invite anyone to steal any or all of these course elements.

Attached figure: Screen shots of experimental videos addressing (clockwise from upper left): baking soda and vinegar reaction; temperature change when evacuating a tank; distinction of reaction kinetics and equilibrium; lower-flammability-limit and ethanol demonstrated with bananas Foster.

a.k.a. sodium bicarbonate

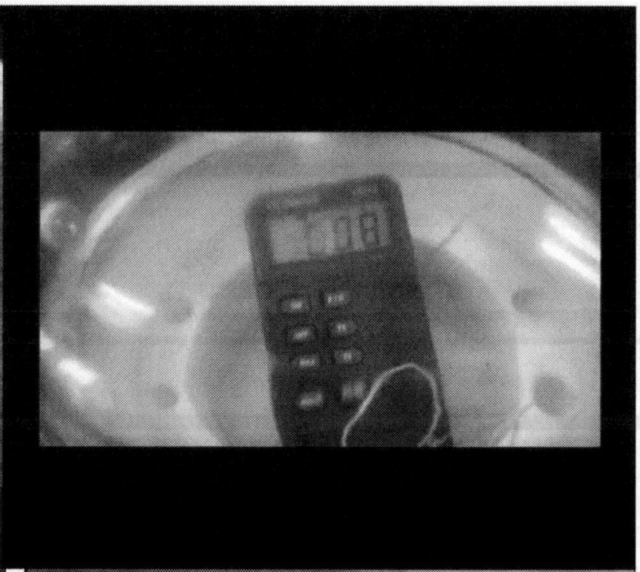

And the answer is Yep! we are

Remember: steel is carbon and iron

632803 Using Social Frameworks and Trusted Influencers to Mediate Inter-Generational Conflicts over Social Justice in the Classroom

Abstract:

Social justice is a critically important concept that should be included as an important criterion in all aspects of society, and chemical engineering is no different. While the push for social justice has long been motivated in the U.S. by our constitutional concepts of equality and equal protection under the law, modern use of cell phone cameras have shown we are woefully inadequate in our execution. Chemical engineering is no different, and efforts to address the disproportionate negative and positive effects of chemical technology on various demographics should be included in our curriculum to produce ethical chemical engineers. To date, most of our engineering students are not as engaged in the social justice movement as many of their counterparts in non-STEM majors are. As such, many chemical engineering educators are not as familiar with the execution of social outreach with their students. Generational differences between many of our Boomer and Gen-X faculty can create significant conflicts with our students, particularly those in the Internet or Z Generation. Drawing on over a decade of work using humor to facilitate educational outreach in the science and engineering fields, we have observed these conflicts first hand. These conflicts are very apparent when using applied humor because the parody inherent in humor can easily trigger such conflicts. Our hypothesis, and those of other researchers, is that the balance between intent and impact is at the root of these conflicts. Despite the fact the both generations are committed to the same social justice agenda, this precarious balance between intent and impact can make bitter enemies out of two parties working for the same goal. Engineering educators need to understand that impact is a much higher priority than intent for many of our current students and conflicts can easily arise that lead to lawsuits and dismissal of educators over differences in perception of intent. There are social/psychological frameworks such as the concept of "Unconditional Positive Regard" that can help faculty avoid these conflicts while discussing social issues in class. We have implemented such a framework in "Humor Genome" class to alleviate conflicts over the controversial nature of humor and parody. This concept is well-characterized by many of our colleagues, including in the writings of Haidt and Lukianoff on the "Coddling of the American Mind". Unfortunately, perception of intent makes these writings suspicious in the eyes of members of Generation Z. Fortunately, similar concepts are espoused by younger social media influencers that are trusted by students so they may be used to address these perception differences in intent and impact to mediate such conflicts. We will discuss the origin of these conflicts in the balance between intent and impact, and show examples from our experiences and from the literature on how conflicts may develop therefrom. We will also discuss how social frameworks and the perspective of trusted influencers may be used to mediate such conflicts should they arise in our efforts to address social justice in the engineering classroom.

621630 Teaching Principles of Biomaterials during the COVID-19 Pandemic with at-Home Inquiry-Based Learning Laboratory Experiments

Abstract:

Introduction: Biomaterials is a multidisciplinary field that has seen a considerable amount of growth over the past several decades, demonstrating a societal need to educate students about the design, engineering, and testing of novel biomaterials. [1] To this end, educators have designed engaging experiments to teach undergraduate students principles of biomaterials. [2], [3] While innovative, these formative hands-on experiences were curtailed due to the outbreak of coronavirus disease 2019 (COVID-19) because many universities transitioned to online learning. As an alternative, we propose providing students with at-home kits that enable them to gain hands-on experience in biomaterials in their own homes. Previously, these types of kits were used in mechanical and electrical engineering education, but few are found for at-home biomaterials experiments. [4], [5] The goal of this educational study is to describe and evaluate three inexpensive, at-home experiments relating to three major classes of biomaterials: ceramics, metals, and polymers, for undergraduate education and STEM outreach.

Methods: Three at-home laboratory experiments were created to teach students principles of Ceramic Toughness, Metal Corrosion and Polymer Stiffness. Materials were purchased from Amazon and shipped directly to students. The total supplies for all three experiments cost less than $100 per student. Additionally, students were taught to use R, ImageJ and Tracker software, which are all open-access software packages commonly used in the biomaterials field. With these tools, laboratory groups completed an inquiry-based pre-laboratory exercise, which helped students brainstorm the exact parameters they planned to test. After in-class discussion, teams were required to submit a written pre-laboratory report that explained the independent variables they were studying, described their methods of quantifiably testing the materials, and calculated a statistically appropriate sample size. After receiving instructor feedback, students had 3-10 weeks to collaboratively complete their experiment and submit a final report.

To assess the learning outcomes of these experiments, pre- and post-tests were given for each experiment. A paired Student's t-test ($p < 0.05$) was used to compare the average pre-/post-test scores. Additionally, an end of semester survey was given to evaluate how the student's perceived the at-home experiments. Descriptive statistics were used to assess the survey results. All surveys were approved by The Cooper Union Institutional Review Board (IRB).

Results: Students successfully completed all three at-home experiments with virtual instruction. For the Ceramic Toughness Experiment, students created various cement slabs by adjusting the porosity of the material. Students then devised an impact testing scheme and quantified the cement shattering patterns using ImageJ. Significant differences in cement breaking patterns were quantified using statistical analysis in R. Twenty students (N=20) participated in the pre/post-test assessment for the Ceramic Toughness Experiment and the average scores significantly increased from 57.5% on the pre-test to 70.0% on the post-test assessment **(Figure 1A)**.

For the Metal Corrosion Experiment, students tested the effects of many biologically-relevant variables that they could devise in their own homes on the corrosion of paperclips. Over the course of 5-10 weeks, Students studied the effects of variables such as galvanization, salinity, and pH. Students were then charged with devising a fatigue testing scheme to assess corrosion differences quantitatively and identify any significant trends using statistical analysis in R. Twenty-one students (N=21) participated in the pre/post-test assessments for the Metal Corrosion Experiment and the average scores increased from 55.6% on the pre-test to 61.1% on the post-test assessment (**Figure 1B**).

For the Polymer Stiffness Experiment, students created physically entangled hydrogel cubes from gelatin solutions of varying concentrations. As an added layer of complexity, students were also asked to create gelatin beads to seed into the cubes and form hydrogel composites. Students explored how altering the concentration of gelatin in the pre-crosslinked polymer solution for the bulk cube or beads impacted compressive properties. Students were challenged to devise their own reliable method to test the compressive strength of the bulk cubes and cube/bead composites. To accurately measure compressive hydrogel displacement, students used Tracker software and identified any statistically significant differences using R. Eighteen students (N=18) participated in the pre/post-test assessments for this laboratory activity and the average scores increased from 55.6% on the pre-test to 64.8% on the post-test assessment with trending significance (**Figure 1C**).

In addition to positive learning gains for all experiments, the results of the end of semester survey demonstrated that students enjoyed the educational experience. Twenty students (n=20) completed the end of semester survey and a majority of students felt that experiments were easily conducted at home in virtual laboratory groups, effectively taught principles of experimental design and data analysis, enhanced course satisfaction and improved learning more than theoretical problem sets would have. Approximately half the students agreed that completing at-home experiments helped develop their comprehension of metal, ceramic, and polymeric biomaterials.

Discussion: This study demonstrated that at-home inquiry-based learning experiments with virtual instruction are an effective way to teach engineering principles during the COVID-19 pandemic. Based on our pre/post-test assessments, we found that students who completed the at-home experiments had positive average learning gains of up to 12.5 points. For comparison, learning gains of approximately 10-15 points were measured for a set of traditional laboratory experiments. [2] Results therefore suggest that the proposed at-home experiments are an alternative to traditional laboratory experiments given the constraints of the COVID-19 pandemic with comparable improvements in comprehension and student engagement.

Based on the end of semester survey results, students welcomed the chance for hands-on experiences while they were unable to have in-person instruction. Other student feedback noted that some students felt that having three inquiry-based learning experiments was overwhelming. The approach of inquiry-based learning, which requires the students to take a more active role in the creation of the laboratory experience, is more time-consuming than traditional teaching approaches. [6] This should be a consideration for educators in the use or creation of other at-home experimental modules. Additionally, the ability to safely and effectively run experiments at-home at a low cost has further implications on expanding access for STEM outreach more broadly.

Outreach is an important activity for increasing the number of students studying STEM and to reach those who are traditionally underrepresented in STEM-based on race, ethnicity, gender, sexual orientation, socioeconomic status, etc. [7]–[9] To help the widespread efforts to promote STEM in these communities, we adapted our Metal Corrosion Experiment to an outreach activity run over two sessions with the Young Eisner Scholars (YES) Program. YES is a program designed to empower middle and high school students from underserved communities. Using pre/post-experiment evaluations, we found that middle school students had significant learning gains and increased positivity towards science after completing our activity. Overall, we hope that these experiments are useful for biomaterials educators and, more broadly, inspire chemical engineering educators to devise creative, at-home experiments for their students due to COVID-19, other disruptions to the traditional, in-person learning environment, and beyond in experiences such as STEM outreach.

References

[1] J. M. Karp, E. A. Friis, K. C. Dee, and H. Winet, "Opinions and trends in biomaterials education: report of a 2003 Society for Biomaterials survey.," *J. Biomed. Mater. Res. A*, vol. 70, no. 1, pp. 1–9, Jul. 2004, doi: 10.1002/jbm.a.30087.

[2] J. Vernengo and K. D. Dahm, "Two challenge-based laboratories for introducing undergraduate students to biomaterials," *Education for Chemical Engineers*, vol. 7, no. 1, pp. e14–e21, Jan. 2012, doi: 10.1016/j.ece.2011.09.002.

[3] S. Houben, G. Quintens, and L. M. Pitet, "Tough hybrid hydrogels adapted to the undergraduate laboratory," *J. Chem. Educ.*, vol. 97, no. 7, pp. 2006–2013, Jul. 2020, doi: 10.1021/acs.jchemed.0c00190.

[4] C.-H. Lee, Y. Liu, M. Moore, X. Ge, and Z. Siddique, "Enhancement of Stay-at-Home Learning for the Biomechanics Laboratory Course During COVID-19 Pandemic," *Biomed. Eng. Education*, Sep. 2020, doi: 10.1007/s43683-020-00025-w.

[5] J. A. Rossiter, S. A. Pope, B. L. Jones, and J. D. Hedengren, "Evaluation and demonstration of take home laboratory kit," *IFAC-PapersOnLine*, vol. 52, no. 9, pp. 56–61, 2019, doi: 10.1016/j.ifacol.2019.08.124.

[6] J. Dewey, "Science as subject-matter and as method," *Sci. Educ.*, vol. 4, no. 4, pp. 391–398, Oct. 1995, doi: 10.1007/BF00487760.

[7] S. Scull and M. Cuthill, "Engaged outreach: using community engagement to facilitate access to higher education for people from low socio-economic backgrounds," *Higher Education Research & Development*, vol. 29, no. 1, pp. 59–74, Feb. 2010, doi: 10.1080/07294360903421368.

[8] C. Riegle-Crumb, B. King, and Y. Irizarry, "Does STEM stand out? examining racial/ethnic gaps in persistence across postsecondary fields," *Educational Researcher*, vol. 48, no. 3, pp. 133–144, Feb. 2019, doi: 10.3102/0013189X19831006.

[9] J. L. Linley, K. A. Renn, and M. R. Woodford, "Examining the ecological systems of lgbtq stem majors," *J. Women Minor. Sci. Eng.*, vol. 24, no. 1, pp. 1–16, 2018, doi: 10.1615/JWomenMinorScienEng.2017018836.

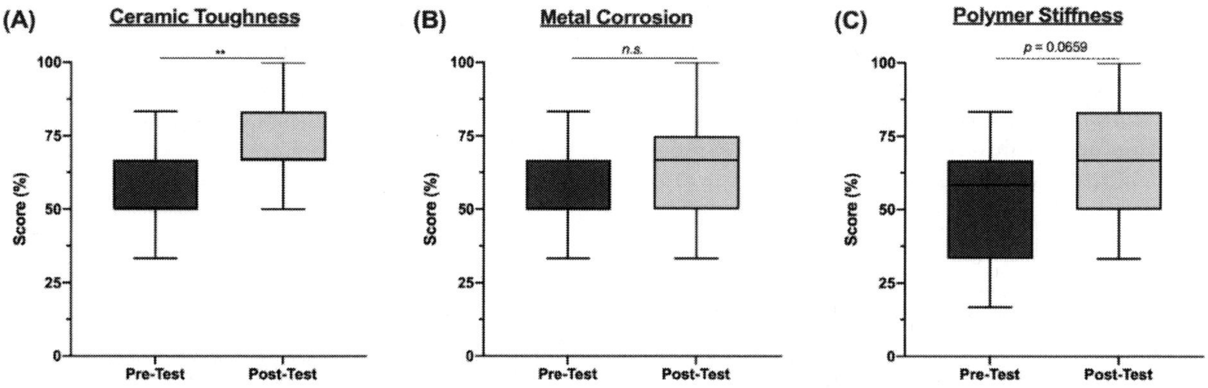

Figure 1. Students scored higher on all post-test evaluations, demonstrating positive learning gains. *Average student pre/post-test scores for (A) Ceramic Toughness, (B) Metal Corrosion and (C) Polymer Stiffness Experiments. n.s = not significant, ** = p<0.01.*

632157 Professional Skill and Attribute Instruction in a Capstone Design Course at an Hispanic-Serving Institution

Abstract:

The senior design course of our chemical engineering curriculum serves as a transition for students completing their undergraduate studies and preparing to enter the engineering workforce. Therefore, the capstone course is a logical place for several professional skills and attributes to be developed by the senior students. The instructor draws on his 25 years of industrial experience for teaching and mentoring communication skills, teamwork, consideration of engineering registration, and the importance of life-long learning. This talk with present the instructor's strategies for developing these skills and attributes in the senior students that will directly benefit them as they enter the professional work force.

Both oral and written communication skills are emphasized in the capstone course, by assigning multiple project reports, technical memoranda, and oral presentations. The extent of communication-based assignments in the course is high for two reasons, namely first because it is the senior design course, and thus project reporting occurs regularly, and second because the course is also the writing intensive course for our chemical engineering curriculum, and thus requires extensive writing requirements to be completed on an individual basis. The expectations for all three of these types of communication assignments are provided by content outlines and grading rubrics provided by the instructor. The students gain experience completing these assignments as both group and individual work, and therefore gain experience akin to a professional environment of the engineer. The group-based senior design project is the course activity the provides a teamwork experience, typically of a longer duration than any previous teamwork experience the students have encountered to date in the university academic career. The students are assessed on their teamwork skills and cooperation by observation from the course instructor, and by completing group contribution ratings of their peers several times a semester.

The students are introduced to the purpose and benefits of professional engineering licensure in our state through a lecture and a visit (face-to-face or webinar) from a representative of the state PE registration board, followed by a topical writing assignment. This instruction is necessary as students consider taking the Fundamentals of Engineering exam at the end of their senior year, and initiate the process of deciding whether to pursue licensure over their early career years. Although the students are all chemical engineers, which is a discipline that has a low level of registration nationally, the instructor emphasizes that our state registration board has recent regulation interpretations stating that chemical engineers working for a consulting or design firm are required to hold a license and sign and seal design documents provided under contract for a refinery or chemical plant operating company. Additionally, the instructor relates his experiences as a registered PE in multiple states over the last 20 years. The concept of life-long learning for those in the engineering profession goes hand-in-hand with professional engineering registration, especially now that most state registration boards also have professional development activities that are required for annual license renewal. Thus, the instructor identifies life-long independent learning as necessary in the professional work environment, and conducts the monitoring of the group senior design projects in a manner that emphasizes the students finding and implementing necessary information on their own.

628991 Holistic, Wrap-Around, Advising Model to Support Student Success and Professional Development

Abstract:

The College of Engineering and Computer Science (ECS) at Syracuse University enrolls about 1500 full-time undergraduate students in 10 degree programs including Chemical Engineering, Biomedical Engineering, Civil Engineering, Environmental Engineering, Mechanical Engineering, Aerospace Engineering, Electrical Engineering, Computer Engineering and Computer Science. Achieving high retention and graduation rates in these disciplines has been a challenge both at Syracuse and nationally. In 2013, our college was awarded a grant through the National Science Foundation (NSF) as part of a special funding focus of the Stem Talent Expansion Program (STEP). The special program was called Graduate 10k+, and the goal was to increase the number of engineering graduates nationally by 10,000 over five years. The title of our project was **E**nhancing the **Cli**mate for **P**ersistence and **S**uccess in **E**ngineering (ECliPSE). Our work focused on enhancing the climate in the classroom, through faculty development in teaching and learning, and outside of the classroom, through development of support systems for students in our college, including an overhaul of our advising system. When the proposal was submitted in 2012 the 1st year retention rate in our college was averaging around 84%, the 2nd year retention rate was averaging around 73%, and the 6 year graduation rate was averaging around 53% compared to 82% for the institution as a whole.

In fall 2013, a small group of first year-advisors in ECS started working more intentionally with students on academic probation by meeting regularly with them throughout the semester to identify the challenges they faced that resulted in their poor academic performance. They provided ongoing skill development and strategies for navigating those challenges and helped them improve their overall performance and experience in ECS. Students who actively engaged in this recovery program saw marked improvement in their performance, many getting cleared from probation at the conclusion of the semester. While they were sometimes reluctant to participate, ultimately, their feedback and experience was generally positive. We began to wonder how implementing this type of intervention might improve the general student experience if executed more proactively, more holistically, and on a larger scale. Over the course of two years, we used our academic recovery program as a model, along with quantitative and qualitative data, to draft a new advising strategy. We were able to garner significant institutional investment in our plans and by the fall of 2018 we had hired six new full-time professional advisors, in addition to two existing staff members, in order to implement our model.

Our advising model takes an expertise approach by providing students with a team of three advisors for all aspects of their success. Their Faculty advisor provides guidance on curriculum, helps demonstrate how a students' coursework connects to their future career aspirations, and provides professional mentoring. The students' Career advisor assists in their career preparation and job readiness skills, helps them secure internships and research opportunities, and provides opportunities for professional networking. And the students' Success advisor supports the students' overall success by helping them identify personal, academic, and professional goals, provides guidance and skill-development for reaching those goals, and provides ongoing, holistic support. This advising team collaborates to provide wrap-around support to our students and the same set of three advisors stays with each student from matriculation to graduation.

Our enhanced advising model was implemented in the Fall of 2018 and after only one year, we saw marked improvement in the student experience in our College. Student engagement with advising in the College significantly increased during the first year of the new model. Our professional advisors (Career and Success) held over 4,500 individual student appointments in 2018-19 and facilitated an increased number of small group workshops. We saw that many first year students engaged with their Success advisors, in particular, much more frequently than required (once a semester), likely due to the connections that students were making with their advisors. In addition, all students are required to meet with their faculty advisor at least once a semester in order to register for the following term.

These connections and the ability of our advisors to intervene early and often in meaningful ways with students translated into a significant improvement in many areas in just the first year of implementation. Our first-year retention rate of 95% tied the highest first year retention rate in the history of our college, which was 11 percentage points higher than the previous year. Additionally, the number of first year students on academic probation after their first semester and first year dropped significantly. The number of first year students on academic probation after their first semester also dropped from 7.6% to 5% and the number of students on academic probation after their first year dropped from 15.3% to 10.9%. In Fall 2018, 76% of students who participated in the academic recovery program were cleared from probation, compared to 45% the previous fall semester. In Spring 2019, 67% of participants were cleared compared to 51% the previous spring semester. In 2018, we saw a significant increase in the number of students reporting that they had accepted a job offer prior to graduation (up 40 percentage points) and we saw a doubling of returning students reporting that they had an internship. This is likely due to increased reporting and the impact of earlier focus on career preparation, all facilitated by the connections built between students and their career advisors. We anticipate continued improvement as student-advisor relationships mature and we continue to assess and develop the model.

An ECS Advising Experience survey was administered in Spring 2020, after students were already studying remotely due to the COVID-19 pandemic. This survey assessed the extent to which it was easy to schedule an appointment with each type of advisor, each advisor took the time to listen to student concerns, each advisor demonstrated knowledge in their area, each advisor provided helpful guidance, and the likelihood that the student would schedule additional appointments (outside of required meetings) with each advisor. Overall, 62.5% of students gave favorable ratings to their career advisor, 73.1% gave favorable ratings to their success advisor, and 80.1% gave favorable ratings to their faculty advisor. High levels of student satisfaction were also reflected in open-ended comments, including:

- *I am very impressed and overall beyond satisfied with my experience with ECS advising. I greatly appreciate the support of my advisors.*
- *My overall experience with advising in ECS has been positive. I have been able to interact with each of [my advisors] a few times throughout the time I have been at SU. I like how easily accessible the advisors are.*
- *I have had a good overall experience with Advising in ECS! My advisors seem to truly care about me and my academic and professional success.*
- *Advising was helpful toward my academic success. Everyone is passionate about helping me succeed...I think the more hands-on approach and activities help students a lot.*

Student retention research shows that many factors contribute to a students' likelihood of persisting and graduating. Some of these are academic advising, social connections, engagement with faculty and staff, and overall student support. Our advising model addresses all of these factors through collaborative efforts to provide our students with wrap-around support from matriculation to graduation. Early data indicates that this model is effective in supporting student success, retention, and career development. Additional data is needed to determine the impact on graduation rates; however, our four and six-year graduation rates have historically tracked with our two-year retention rates, so it is reasonable to expect that graduation rates will increase as first and second year retention rates increase. We recognize that this advising model is more resource intensive than what we were previously able to achieve without institutional investment. But the return on the investment has been tremendous and these high impact practices have already had dividends in terms of retention rates, probation rates, career outcomes and in enhancing the overall climate for persistence and success in our college.

627840 "10.801: Project Management and Problem Solving in Academia and Industry" — a New Course for MIT Graduate Students in Chemical Engineering Practice

Abstract:

At the core of MIT's long-established Master of Science in Chemical Engineering Practice (M.S.CEP) program are four one-month projects at the field stations under the direction of resident MIT Practice School Station Directors. Over the years, the directors have observed that students often have difficult absorbing certain educational material on project management and problem solving while simultaneously attempting to successfully complete their projects. Hence, we identified an opportunity to teach the students these "soft skills" back on campus in a brand-new course, 10.801. The course was put together during the year of 2020 by a team of eight instructors consisting of faculty, Station Directors, and industry professionals. We organized a pilot run of the course in January 2021 during the Independent Activities Period (IAP) at MIT, with 13 students attending remotely. Topics covered included: career development, project management, library resources, integrated problem solving, project economics, safety, time management, and professional behavior. We hope to accommodate a greater number of students in the second run of 10.801 in January 2022 by moving toward a blended instruction paradigm consisting of both asynchronous and synchronous learning activities. The asynchronous contents produced could also eventually lead to a Massive Open Online Course for broader use by the chemical engineering community.

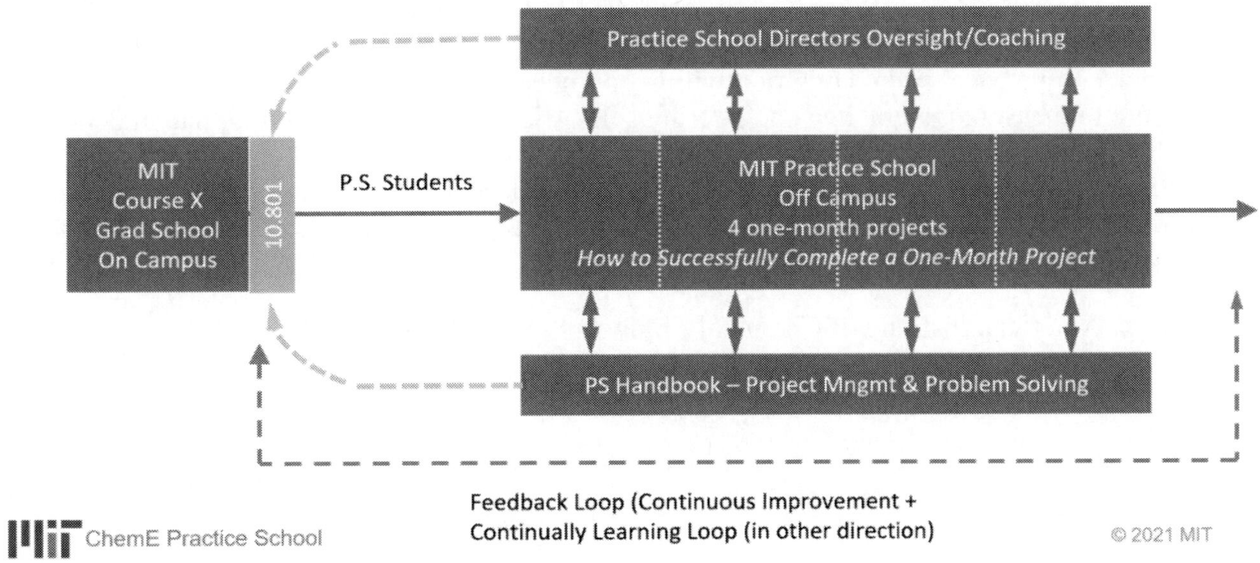

626122 Relationship between Students' Engineering Identity and Self-Efficacy and Participation in an Engineering Education Enrichment Program

Abstract:

The development of engineering skills through engagement in co-curricular learning strategies is an important contributing factor for preparing engineering students for the global workplace. There is also evidence linking engineering identity and self-efficacy to educational and professional achievements. While the effects of co-curricular and extracurricular activities and engineering identity on academic outcomes have been studied, direct measurements relating engineering identity and self-efficacy and choosing to participate in activities that prepare students for effective professional careers are not available.

The College of Engineering at New Mexico State University (NMSU) has developed an Engineering Education Enrichment (e3 Initiative) program funded by the National Science Foundation to help students to develop professional skills through industry-guided design projects, industry-recognized certifications, and entrepreneurship training. To examine whether students with higher engineering identity and self-efficacy are more likely to participate in the e3 Initiative, a survey was administered in Fall 2020 to NMSU undergraduate students from all engineering majors who were invited to participate in the e3 Initiative. Survey instruments had been developed and validated in previous research to measure identity based on three key constructs: self-beliefs of their interest, performance/competence, and recognition within engineering. Several items were also adopted from the literature to measure general and engineering self-efficacy. We expect to find that students who have high engineering identity and engineering self-efficacy will be more likely to participate in the e3 Initiative program. A better understanding of how engineering identity and self-efficacy are linked to co-curricular activities can help us to better support the academic success of all engineering students.

621797 Student-Driven Career-Focused Development of Design Engineering Skills

Abstract:

Forward-thinking undergraduate students seek extracurricular activities and membership in professional societies to develop knowledge and skill sets that complement their core coursework with the goal of increasing their marketability and employment prospects. Engineering students with industrial and internship experience have much they can offer their fellow classmates regarding industry-relevant knowledge and career-oriented skill sets. A student organized and student taught CHEM-E-CAD course was piloted in Fall 2019 under the supervision of a faculty mentor through the chemical engineering program at a large research institution with the purpose of augmenting the knowledge and skill sets of undergraduate (chemical) engineering students in the areas of CAD software, 3D printing, circuit design and assembly, and industrial automation. These topics involve modern engineering tools that are not typically included within the scope of the standard chemical engineering curriculum but are nonetheless valuable to many companies from which the students may wish to gain employment. This course has continued to be offered each quarter since the pilot offering for a total of 6 unique offerings. Each offering has been structured such that continuing students can learn something new and new students can learn the basics of that quarter's topics. The CHEM-E-CAD course was led during the first year (3 offerings) by a student instructor who leveraged his industrial experience to share his knowledge in these areas. During the second year, the course was led by a group of students who benefitted from the group during the first year and desired to share their knowledge with the incoming group of students. Students enrolled in the course attended weekly meetings consisting of presentations by the student instructors, group discussion of concepts, and workshop/lab sessions. Course activities were geared towards encouraging each student to create virtual and physical products that could be included in his or her professional portfolio. Each student needed to complete a predetermined number of assignments in order to receive a "pass" for this 1-unit pass/no pass course. Within the first two years of CHEM-E-CAD, over 120 individual students have benefitted from being a part of this program. After each of the six offerings, participants rated their perceived achievement of course outcomes and provided feedback, which has been overwhelmingly positive, to the student instructors and faculty mentor. During this presentation, the experiences and perspectives of the student leaders will also be shared.

631596 The Prevalence of Imposter Phenomenon in Undergraduate and Graduate Chemical Engineers

Abstract:

Imposter phenomenon (IP) was first characterized in high achieving women during the late 1970s but has since been found across a wide variety of populations and identities. For instance, higher levels of IP have been documented in high achievers, women and underrepresented populations. Recently, a Science letter was published that implicated IP as a potential threat to diversity in STEM. In the engineering context, Richard Felder first drew attention to IP in engineering students through an article published in Chemical Engineering Education in 1988. Since then, there have been many opinion pieces written about the presence of IP in the engineering population. That being said, there has been little research to quantify the prevalence of IP specifically within this student population. In this study, results will be presented on the prevalence of IP in undergraduate and graduate chemical engineering students at the University of Kentucky. The study will use the Clance Imposter Phenomenon Scale to measure IP levels in students across varying age of study. Results will be used to further motivate the study of IP specifically within the engineering student population at both the undergraduate and graduate school levels.

632553 How to Make a Data Science Curriculum Embrace Engineering Domains and Vice Versa

Abstract:

The world's attention on big data, data science, and machine learning has been triggered by successful applications in recommendation systems, natural language processing, computer vision, image classification, autonomous systems and processes, and social media, etc. The technological advances are transforming traditional industries, manufacturing, and business operations. Compared to the successful industrial applications, the immersion of data science education into science and engineering curricula is just at its beginning and fast developing from graduate to undergraduate programs (Hicks and Irizarry, 2018; Braatz et al., 2019). New schools, departments, minors, concentrations, and modules of data science have been established worldwide in the recent few years.

A revolutionary impact of data science on science discoveries is the addition of data analytics to the three pillars of scientific research, i.e., theory, experiment, and computing, especially where models are not well established, such as the design of new material molecules with desired properties. On the other hand, industries are seeing the next industrial revolution that unleashes values in massive data generated from operations, manufacturing, and service systems.

However, due to the diverse disciplines of engineering and the even more diverse sets of analytical methods in data science, a critical challenge is how to create a curriculum that will equip the next generation workforce with the highly needed skills in data science and specialization in an application domain. This approach of data science plus a domain (i.e., DS+X) was adopted at the City University of Hong Kong to establish the first stand-alone School of Data Science in Asia. In this talk, we will introduce the curriculum design of the School of Data Science which embraces engineering and science domains, including energy, environment, smart manufacturing, smart cities, social media, and health analytics. The relevance to chemical engineering will be focused on. Conversely, we will discuss how chemical engineering education can be enriched with data science courses, modules, and even a minor program in data science. While data science applies to many areas in chemical engineering, it seems that it would be a natural enhancement of process systems engineering. It is anticipated that data scientists with a domain specialization will be in high demand.

References

Stephanie C. Hicks and Rafael A. Irizarry (2018). A Guide to Teaching Data Science, The American Statistician, 72:4, 382-391, DOI: 10.1080/00031305.2017.1356747.

Richard Braatz, Lloyd Colegrove, Sharon Glotzer, Curtis Martin, Joe Qin (2019). Data Science Education in Chemical Engineering, Panel Discussion Chaired by Leo Chiang, AIChE Annual Meeting, November 11, 2019, Orlando, FL.

623470 Interactive Software for Teaching Multivariable Data Analytics

Abstract:

Latent variable methods such as partial least squares and principal component analysis are among the most widely applied data analytics tools for applications to chemical and biological processes. It can be challenging to teach these multivariable statistical methods in a way that the students gain enough conceptual understanding to be able to consistently apply these methods effectively in practice, especially considering that many chemical engineering programs do not require that their students take courses in linear algebra or applied statistics. The result is that many practicing chemical engineers use alternative methods instead, such as correlating a peak absorbance to each concentration when calibrating a spectral measurement, which can produce much lower accuracy models. Alternatively, the latent variable methods are often used in a black-box manner, without having enough understanding to know when a situation occurs in which the methods should not be applied, or how to revise the way that data are fed to the method to work around some imperfection in the data such as sensor bias.

Especially for students who have not had linear algebra or applied statistics training, the mathematical details of latent variable methods can be challenging to understand. The tendency will be for students to treat each method as a black box. After all, many software packages are available for applying latent variable methods, and it might seem that a more profound understanding is not needed for problem-solving. What is more, often multiple methods are applied to a problem at hand, and decisions for algorithms are made solely based on preliminary results. This approach makes it challenging to explain and interpret the models that are creating by these methods, in particular, in relating the models that are generated with the chemistry/biology occurring in the process, to reconcile the data analytics results with domain knowledge. It is the synthesis of domain knowledge with data analytics that is the added value of a well-trained chemical engineer and is likely to result in the best chemical engineering solutions for the particular problem at hand. Furthermore, preliminary black-box results can lead to choosing overly complicated models that overfit the data. Advancing the understanding and intuition on latent variable methods is needed to avoid overfitting for some types of biased data and ultimately assure model interpretability, leading to higher value, acceptance, and applicability.

This presentation describes software and examples that were developed to train students to achieve a deep understanding of latent variable methods [1],[2] and the related machine learning methods of lasso [3] and elastic net [4] (for the remainder of this abstract, these methods will be collectively referred to as latent variable methods, although some of these methods are more associated with the machine learning community). The graphical user interface was designed for the explicit purpose of teaching undergraduate and graduate students, which is a distinguishing feature from the graphical user interfaces in existing chemometrics software packages which are focused on just applying a method to a dataset. The software takes the perspective of the optimization being solved, so that the students can gain an understanding of the relationship between the latent variable method that is selected and the results that are produced.

This tool, referred to as Latent Variable Demonstrator (LAVADE), compares a wide range of latent variable regression techniques with traditional regression techniques on carefully designed examples. The examples are designed to be easy to understand, and various options to customize the problem are available to learn exactly how the different algorithms approach the model construction. Perturbing the signal step by a step with more noise fosters an understanding of how the different methods deal with noise. Furthermore, the tool allows the student to play and compete with the algorithms, making it exciting to gather knowledge and intuition to explain the algorithms' behavior on real-world problems.

References:

1. Leo H. Chiang, Evan L. Russell, and Richard D. Braatz. Fault Detection and Diagnosis in Industrial

Systems. London, UK: Springer Verlag, 2000.

2. K.V. Mardia. Multivariate Analysis. London, UK: Academic Press, 2003.

3. Robert Tibshirani. "Regression Shrinkage and Selection Via the Lasso." Journal of the Royal Statistical Society: Series B (Methodological), 58(1), pp. 267–288, 1996.

4. Hui Zou and Trevor Hastie. "Regularization and variable selection via the elastic net." Journal of the Royal Statistical Society. Series B: Statistical Methodology, 67(2), pp. 301–320, 2005.

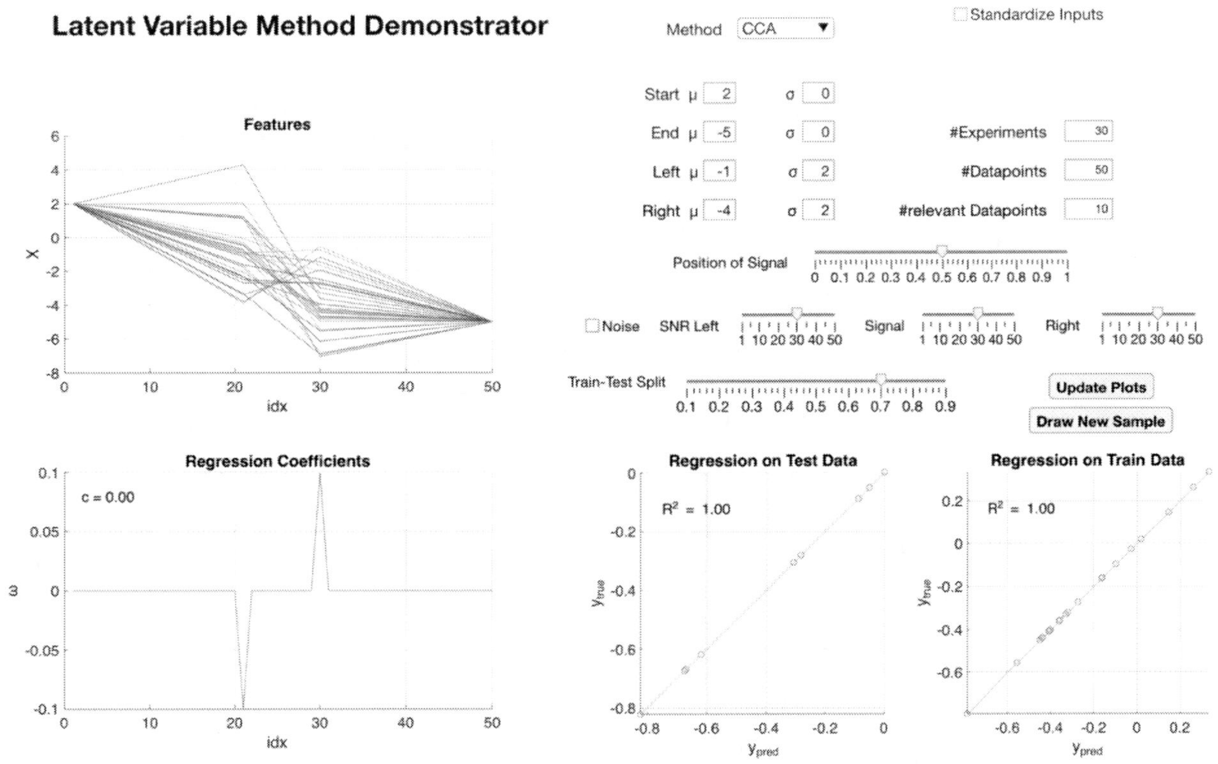

632834 Chemical Engineering Students' Perception of and Attitudes Towards Data Science

Abstract:

The future of chemical engineering (CHE) is changing relative to the impact of novel analytics and large-scale computations. However, CHE curricula have not kept pace with the increasingly complex problems that accompany ongoing technological advancements. Hence, there is greater importance in developing skills in data science as the use of technology continues to escalate. Understanding the attitudes that CHE students have towards Python, Excel/VBA, statistics, Matlab, and other data science instruments will allow universities to improve their method of administering skills and knowledge that are vital to overcoming challenges faced in the industry.

This research utilized a survey to determine how and what specific factors influence the perceptions 100 chemical engineering students at the University of Illinois at Chicago (UIC) have surrounding this matter. In addition to the computing teacher's demographics, the analysis conducted also investigated how the students' demographic characteristics, skills in computing, class standing, and industry and research experiences play a role in their viewpoints.

The research team collected both quantitative and qualitative feedback from the participants' survey evaluations. With the use of a Likert scale of five or ten, numerical responses were gathered from questions geared towards the participants determining the degree to which a certain statement pertains to them. In addition, participants were allowed to further explain the reasoning for their stance on a particular topic. Verbal responses from these entries enabled the research team to have further insight into how specific variables have a greater influence on one group of participants than another. From the data collected, the research team viewed and cross-referenced the trends of the responses to draw conclusions on whether, and how, the criteria previously mentioned play a role in students' viewpoints towards learning data science.

This paper will provide a detailed explanation of how the survey was developed and analyzed, data distribution and findings, and methods in which universities can better develop and integrate future data science programs into their curricula.

630729 Development of Static and Dynamic Simulation-Based Active-Learning Modules for Chemical Engineering Curriculum

Abstract:

As technology and specifically dynamic simulators become commonplace in engineering programs, engineering educators have a great opportunity to evolve the way they approach engineering education. Most students are familiar with the traditional lecture structure commonly used in engineering classrooms across the United States. In this structure, a new concept is introduced to the students, a derivation is performed to create a useful tool for solving problems, an example problem is then proposed, and the example problem is solved using the derived formula. Although this structure feels like a logical methodology for teaching a new concept and reinforcing it through application via problem-solving, there is little evidence in literature to support this approach. In a study done by Prince, et. al. involving 344 undergraduate chemical engineering students in a heat transfer course, they found that this traditional approach to teaching STEM (science, technology, engineering, math) generally does not develop conceptual understanding of the material [1]. The education literature seems to confirm this finding as well [2]–[4].One of the contributing factors to this is the lack of active learning involved. Although it may feel like solving a real-world problem as described earlier should cover this, it rarely engages or challenges students. Rather than students learning the concepts to use in the problem-solving process, they learn the procedures/algorithm for solving a very narrow family of problems.

Another major factor in the success and failure of students in engineering courses is the difference between "sensors" and "intuitors." According to the literature, sensors are students who prefer the practical side of learning whereas the intuitor enjoys exploring abstract concepts. If the vast majority of engineering teaching relies on this derivation-based approach to teaching, then it will disproportionately benefit intuitors and sensors will be left behind [5]. This means, there is a gap in engineering pedagogy when it comes to active learning techniques that supplement this practical side to STEM education that can complement the theoretical side. With technology becoming more common in engineering course work, dynamic simulators offer a great opportunity for improving engineering pedagogy in this area.

In this work, the authors have identified ten courses common to most chemical engineering undergraduate programs in the United States. For each course, four active-learning activities are developed to help reinforce concepts taught in these courses. To do this, the activities are developed into individual modules consisting of background, the activity itself, and a debrief. Literature has found that students attention in the traditional lecture begins to decrease around ten minutes into the lecture [4]. Because of this, the background sections are limited to 15 minutes or less. The activity consists of either solving problems by hand and validating them in the dynamic simulator or by solving the problem by using the only simulator. The simulator employed here is AVEVA Process Simulation, which provides both options of simulation or model writing roles, being an ideal fit for different equation-oriented and simulation-based courses. Lastly, suggested debrief questions are provided to help reinforce the concepts learned through the activity. These modules are then packaged into one teaching reference guide.

The development of such a guide will be a useful tool for faculty interested in using the technology that is becoming more common in engineering programs in a way that promotes active learning. Such a resource greatly reduces the effort required of faculty and creates a better learning environment for students. This ultimately helps faculty become more effective teachers and students to become better learners.

References

[1] M. J. Prince, M. A. S. Vigeant, and K. E. K. Nottis, "Assessing the prevalence and persistence of engineering students' misconceptions in heat transfer," *J. Eng. Educ.*, vol. 101, no. 3, pp. 412–438, 2012.

[2] S. A. Ambrose, M. W. Bridges, M. DiPietro, M. C. Lovett, and M. K. Norman, *How Learning Works: Seven Research-Based Principles for Smart Teaching*. San Francisco, CA: Jossey-Bass, 2010.

[3] A. K. Taylor and P. Kowalski, "Student misconceptions: Where do they come from and what can we do?," in *Applying science of learning in education: Infusing psychological science into the curriculum*, Washington, DC, US: Society for the Teaching of Psychology, 2014, pp. 259–273.

[4] R. M. Felder and R. Brent, *Teaching and Learning STEM: A Practical Guide*. Hoboken, UNITED STATES: John Wiley & Sons, Incorporated, 2016.

[5] R. M. Felder, G. N. Felder, and E. J. Dietz, "The Effects of Personality Type on Engineering Student Performance and Attitudes," *J. Eng. Educ.*, vol. 91, no. 1, pp. 3–17, 2002, doi: https://doi.org/10.1002/j.2168-9830.2002.tb00667.x.

629030 Introducing Students to Open-Source Partial Differential Equation Solver Codes in Python

Abstract:

Partial differential equations (PDE) are ubiquitous in many engineering fields and are one of the most complex mathematical topics that engineering students will encounter during their education. Despite the fact that numerical methods are often required for effective solution of PDE models for research and industrial applications, undergraduate students typically will have very limited exposure to relevant numerical methods and/or the use of PDE solver codes. This situation arises for a multitude of reasons such as the availability of space in the curriculum and/or access to suitable resources. Whilst addressing curriculum constraints is a challenging task, the availability of high quality resources suitable for teaching purposes can be well addressed. To that end, we believe that recent releases of multiple PDE solver codes native to the Python language offer exciting opportunities for educators and students. These solver codes are typically open-source in nature which makes them readily available. They are also easy to use as Python is a comparatively easier language for students to work with and are highly scalable, ofteni finding use in many research projects and publications. As a proof of concept, we have developed a short course (6 hours of contact time) using the solver code FiPy that we hope to integrate into the teaching of an advanced engineering mathematics course for 2nd year undergraduate students. With appropriate scaffolding, students were able to understand and solve a variety of problems including the Cahn-Hilliard equation. Student feedback indicated that they broadly found the course engaging and useful and that the visualization capabilities greatly facilitated their understanding of key concepts. All the course material is currently available on a public Github repository and we hope to implement additional exercises and use other solver codes in the future.

622383 Using Templates for Making Programming Languages More User Friendly

Abstract:

In a survey that has been carried out recently by Hesketh et al. (2020) they found that: "... ***just over half*** of faculty believe that graduates are well prepared to use computer software and programming to solve engineering problems." They also pointed out that a possible reason for the low integration of software use is the perceived barriers for its use by faculty. In response to the question on what the barriers for faculty are for not using computer aids, the most common response was that faculty were not trained in these computer tools, and that faculty do not have the time required to learn how to use these tools. That survey indicates that many faculties did not use these software programs as undergraduates and currently do not use the software in their research, so there is little incentive to learn programming.

However, our experience has shown that it is possible to create software packages that are easy to learn and apply by both the students and the faculty. The POLYMATH package was introduced by Shacham and Cutlip about 30 years ago (e. g., Shacham and Cutlip, 1992) and it is still being successfully used, both in education (e.g., Valdivia et al. 2021) and scientific research (e.g., Anitha and Garlapati, 2020). POLYMATH is easy to use because it requires the user to input only the information related to the problem which is being solved.

For example, both Valdivia et al. (2021) and Anitha and Garlapati, (2020) use POLYMATH for carrying out nonlinear regression to fit a nonlinear model to experimental data. In such case the user needs to input only the experimental data, the equation representing the nonlinear model and initial estimates for the model's parameter values. Using this input, the program calculates and displays the optimal parameter values, some statistical indicators of the goodness of the fit, and produces various plots that demonstrate the goodness of the fit between the experimental and calculated data.

This arrangement is very convenient for scientists, who are interested in obtaining the desired results with minimal programming effort. However, it is inappropriate for students learning programming. POLYMATH is able to output a MATLAB script file, but if they don't have access to POLYMATH then they will need to write this file. If they used POLYMATH they could follow the logic from entering equations in POLYMATH to constructing a MATLAB script file. Furthermore, to keep the use of POLYMATH simple, its use is limited to Single-Model Single-Algorithm (SMSA) problems. Solution of complex Multiple-Model and/or Multiple-Algorithm (MMMA) problems require the use of more versatile programming languages. More detailed discussion of this issue can be found, for example, in the review paper by Shacham et al. (2009).

` Based on these observations, we are attempting to develop computing tools that allow easy and efficient numerical problem solving, however also enable studying computer programming at both basic and advanced levels. The selected tool for this purpose is a template, implemented, for example, in MATLAB, where the user is required to enter only problem specific information. The MATLAB program solves the problem and displays the results in numerical and graphical forms. But unlike in the case of POLYMATH, the entire code of the program is visible to the user and can be modified by him, as necessary.

At this stage we are developing templates for solving nonlinear algebraic equations (NLE), ordinary differential equations (ODE), and linear, polynomial, and nonlinear regression, for MATLAB and Python. This is supported by the finding of Hesketh et al. (2020) that the predominant programming languages taught in "computing in chemical engineering" courses are MATLAB (78%) followed by Python (31%). The templates enable scientists to solve specific types of problems without learning programming, thus the effort required for problem solution is similar to the effort required when using POLYMATH for the same purpose.

For students learning computer programming the templates are provided at the start of the course. First the students are assigned exercises that can be solved by the appropriate template just by providing the problem specific information. Consider, for example the problem presented by Shacham et al. (2003). In this exercise the molar volume (V), the compressibility factor, the isothermal enthalpy and entropy departures, and the fugacity coefficients are calculated and plotted for water. In this example, the single-NLE solver template can calculate the molar volume at one temperature (T) and one pressure (P) values. The user provides only the function that calculates $f(V)$ for a specified value of V, and an initial estimate for the molar volume value. By solving similar examples with the appropriate templates, the students experience the importance of the programming in problem solving. In a later, more advanced stage of the course, the students are required to modify the template-based solution, for this example, to enable the calculation, display, and plot of all the additional variables (enthalpy and entropy departures, etc.) studying the effects of the temperature and pressure variations and display all the pertinent values.

In the extended abstract and the conference presentation we present in detail an example of the dual template use. It demonstrates the potential of the new approach to facilitate process computations and learning programming via gradual enhancement of the template program to enable solving problems of higher complexity level. This approach will be demonstrated by solving boundary value (a single-model multiple-algorithm type) problems. Preliminary results of the classroom testing of the template-based programming study will also be reported.

References

1. Anitha, N. and C. Garlapati, "Solubility model correlations of some substituted phenol compounds in supercritical carbon dioxide", AIP Conference Proceedings 2270, 040013 (2020); https://doi.org/10.1063/5.0019510, Published Online: 02 November 2020.
2. Hesketh, R. P., Grover, M., & Silverstein, D. L. (2020). CACHE/ASEE survey on computing in chemical engineering. *ASEE Annual Conference and Exposition, Conference Proceedings, 2020-June*, [293].
3. Shacham, M. and M. B. Cutlip, "POLYMATH: General, User-Friendly Numerical Computation Package", paper 226a, presented at the AIChE 1992 Annual Meeting, Miami Beach, Florida, Nov. 1-6, 1992.
4. Shacham, M., Brauner, N. and M. B. Cutlip, "An Exercise for Practicing Programming in the ChE Curriculum--Calculation of Thermodynamic Properties Using the Redlich-Kwong Equation of State.", Chem. Eng. Educ., 27(2), 148 (2003)
5. Shacham, M., Cutlip, M. B. and N. Brauner, "From Numerical Problem Solving to Model Based Experimentation – Incorporating Computer Based Tools of Various Scales into the ChE Curriculum", Chem. Eng. Educ, 43(4), 315 (2009)
6. Valdivia, A. E. O., Osornio, C. M., and Y. M. V.-Rodríguez, "PBL with the Application of Multiple and Nonlinear Linear Regression in Chemical Kinetics and Catalysis", American Journal of Educational Research, 9(1) 11-17 (2021),

629222 Matlab Based Applications As Accessible and Interactive Educational Modules to Advance Spectroscopic Understanding

Abstract:

Spectroscopic techniques are unquestionably valuable for understanding structural and compositional changes in complex chemical, biochemical and biological systems. Consequently, spectroscopic techniques are increasingly incorporated to undergraduate/graduate curricula to prepare the next generation of engineers for successful careers in industry and academia where applications include reaction monitoring, quality control, and process control etc. Spectroscopic pedagogy excels in theory and interpretation of spectra and some educators are shifting towards programming-based lessons to introduce certain topics. The interactivity and exposure of computer science in the field of chemistry are advantageous in a future scientist's career.

Among the challenges in traditional spectroscopic pedagogy, two stand out: the practical aspect of collecting meaningful and reproducible data along with the relevant and necessary pre-processing prior to any interpretation. Lab-based sessions are expensive and certain techniques may be prohibitively long or sensitive for students to broach topics such as instrumental parameters or background drifts. Through programming-based modules developed for education, these topics can be introduced earlier to make higher education level students aware of the reality of spectroscopic techniques in addition to the theory. We have developed an extensible MATLAB program which allows students to explore Raman instrument parameters to maximize signal to noise ratio while minimizing collection time. Current efforts are focused in extending a MATLAB module to interactively showcase preprocessing techniques with the intent to feed into chemometric interpretation and modelling. Spectral preprocessing is applied in nearly every publication, but justification is often neglected despite the negative impact of incorrect choices in interpretation and modelling. Finally, the benefits and pitfalls of developing MATLAB for the chemical engineering educator is discussed.

Sustainable Bioeconomy for Arid Regions (SBAR): A coordinated approach to establishing bioeconomies

Abstract

SBAR is a Coordinated Agricultural Project or CAP funded by the National Institute of Food and Agriculture, USDA. The overall goal is to determine the types of crops that require low amounts of water, are natural to the arid southwest, and can be used to strengthen the bioeconomy in the region. Guayule is the major crop being investigated. SBAR combines the expertise of universities (University of Arizona, Colorado School of Mines, Colorado State University, and New Mexico State University), USDA/ARS and Bridgestone Americas, Inc. Research accomplishments include: development of agronomic practices to reduce water requirements; identification of select germplasm producing higher yields at lower irrigation rates; new field models for irrigation application; understanding soil health through microbiome analysis; identification of guayule secondary metabolites that will potentially result in specialty co-products; and the distillation and characterization of guayule resin to generate market commodity co-products, such as terpenes, bio-based adhesives and insect repellents. The team has developed and validated an integrated life cycle assessment model for guayule, including an integrated economic and environmental impact assessment tool, and developed a model for farm water need/use. Metrics for social sustainability in alignment with UN Sustainable Development Goals have been identified, which inform future research needs for critical social sustainability factors in the Southwest. An overview of the project and key results will be presented.

627049 High-Throughput Measurement of the Human RED Blood Cell Shear Modulus Distribution As a Function of PO_2

Abstract:

One of the fundamental questions around the human circulatory system is to determine how blood flow is regulated so that blood can be distributed to the right place when and where it is needed. For example, neural activity has to be accompanied by an increase in local blood flow to satisfy increased glucose and oxygen demand–the partnership between neural activity and local blood flow is termed *functional hyperemia*. In landmark work, J. Wan and his colleagues[1] very recently showed that cerebral functional hyperemia is initiated in the capillaries with ***RBCs acting as oxygen sensors to regulate cerebral blood flow***. In this case, the interactions between deoxygenated hemoglobin (deoxyHb) and band 3 protein in the RBC membrane are the molecular switch that responds to local PO_2 changes and controls RBC deformability and capillary blood cell velocity. We believe that this phenomena, whose study is in its nascent stages, may directly bear on COVID19 and the development of this and other diseases. This follows directly, since it is now known that SARS-COV-2 attacks hemoglobin directly and hypoxia is a generic symptom of COVID19[2]. Indirectly, decreased flexibility of RBCs has already been noted as a biomarker for sepsis, with the cell membrane altered by the attacking micro-organisms. Note in this context, that sepsis creates a coagulopathy involving bleeding, while SARS-COV-2 creates a coagulopathy associated with thrombosis—the latter of which is currently a subject of intense study[3].

RBC structure and deformability are largely governed by the ***membrane shear modulus***. State-of-the-art methods to measure the shear modulus of RBCs are not high-throughput and, microfluidic platforms for high-throughput measurements of RBC mechanical properties have not yet enabled measurement of the shear modulus. These limitations challenge the development of diagnostic devices based on RBC shear modulus biomarkers. In this talk, we will review the development of our high-throughput microfluidic platform[4], coupled with high-fidelity simulations to address this significant gap in technology. In contrast with existing technologies, this approach allows us to measure the shear modulus of individual RBCs and generate shear modulus distributions (for a given individual or multiple individuals) including measurements of thousands of cells in a few seconds of experimental data acquisition. We demonstrate that our platform provides mean values of the modulus that are in quantitative agreement with other low throughput measurement techniques in the literature (e.g optical tweezers), but that the distribution for each donor follows a broad lognormal distribution.

We will then discuss the modification of our platform to make similar measurements of an individual's shear modulus distribution but at controlled PO_2 saturation levels in the blood. Thus, our microfluidic chip is modified to include gas channels that run along the blood flow channel. The PO_2 titration in the gas channels is accurately controlled such that for each individual blood sample measurements of the shear modulus are made at the equilibrium PO_2 of the gas channel. The result yields accurate and independent measures of the correlation of RBC shear modulus with PO_2 present in healthy individuals for the first time.

[1] Zhou *et al*., **Oxygen tension–mediated erythrocyte membrane interactions regulate cerebral capillary hyperemia,** *Sci. Adv.* 2019; **5** : eaaw4466 29 May 2019; [2] Cavezzi et al., **COVID-19: hemoglobin, iron, and hypoxia beyond inflammation. A narrative review**; *Clin Pract.* 2020 May 19; **10**(2): 1271.doi: 10.4081/cp.2020.1271; [3] Connors, J. And J. Levy, Blood, 135(23) pp.2033-2040 [4] Saadat et al., High-throughput measurement of an individual's red blood cell shear modulus distribution, *Lab on a Chip* , **20**, pp. 2927-2936 (2020).

638189 Interactive Self-Study Modules for Chemical Engineering Education

Abstract:

Retention of diverse students in STEM disciplines has emerged as a major goal for educational research and tool development. There is a major need for modular resources that can be used to help students catch up on key concepts or as enabling components of active learning approaches that enhance student learning. In this presentation, we will introduce the use of interactive self-study modules (ISSMs) and discuss our investigations into how these modules can be designed to assist student self-directed learning. These web-based ISSMs are used either as part of a flipped classroom approach or for students to use on their own; each takes approximately one hour to complete.
These modules incorporate approaches that have been shown to increase student learning. The ISSMs contain the following elements: (i) an overview that describes the goals of the module and lists pre-requisites; (ii) formative assessment using conceptual questions (ConcepTests) and problems; (iii) short introductory screencasts with embedded questions; (iv) lists of important equations; (v) interactive simulations with questions asking students to predict system behavior; (vi) an interactive quiz-yourself simulation that requires student input in a complex problem solution; (vii) example problems with screencast solutions; (viii) ConcepTests for summative assessment; and (ix) a summary with problem answers and a list of what students should be able to do after completing the module. The ISSMs, which are freely available on learncheme.com, are part of an overall approach to incorporate frequent self-testing (active retrieval) and active learning as evidence-based teaching methods.

624344 Combining Scientific Rigor and Outside-the-Box Thinking – Perspectives of a Former Advisee

Abstract:

The Orr-Sommerfeld equation meets synthetic biology. Particle flocculation, when positioned in the proper flow field, solves a genetic problem. More appropriately, scientific rigor, when mixed with curiosity and creative thought, leads to societal solutions. When these are components of an education enterprise, contributions are amplified. This presentation builds on inspirations from Rob Davis and correspondingly, will highlight efforts from our laboratory that have been based on first principles, stretched our imaginations, and expanded disciplinary boundaries. Larvae that glow. Bacteria that talk. Redox information channels that connect biology to electronics.

624487 Progress in Therapeutic Biologic Production Technologies and Future Challenges

Abstract:

Over the last 30 years, a series of advances in protein production technologies, often derived by collaboration at the interfaces of engineering, chemistry and biology, have led to improvements of several orders of magnitude in process efficiency enabling a wave of therapeutic products. In this presentation, several examples of these advances will be shared along with important remaining challenges for the future including implications on educating the engineers and scientists to solve these problems.

619913 Polymers and Light: Towards Advanced, Photoresponsive Materials

Abstract:

Light induced reactions including photopolymerizations have been explored and utilized since the time of the ancient Egyptians; however, development of new photoresponsive materials, methodologies and applications continues at an ever more rapid pace. Light induced reactions are ubiquitous because they are far more energy efficient than their thermal counterparts, are typically performed in a solventless manner that is more environmentally compatible, occur rapidly at ambient conditions, and they can be controlled in both time and space.

Here, we will focus on two distinct vignettes related to our work on light induced reactions including the development of a bistable material that exhibits controlled, photoinduced transitions from a viscoelasetic fluid to an elastic solid upon exposure to light. This work on covalent adaptable networks (CANs) incorporate dynamic covalent bonds as crosslinks within the network that respond to light to (de)activate the dynamic bond character. These materials are able to change shape and stress-state when exposed to light, facilitating application as novel optical and actuating materials. Using thiol-thioester exchange chemistry, we will discuss the formation of a material that is capable of undergoing recycling, healing, and enhanced toughness of materials based on the photo(de)activated dynamic covalent chemistry. Ultimately, the potential for CANs-based materials to impact numerous materials applications will be presented in light of their distinctive array of material properties.

Secondly, we will present a novel approach to the fabrication of semicrystalline thermoplastic polymers via photopolymerization for potential use in additive manufacturing. Thermoplastic materials produced from photopolymerization processes are notoriously difficult to produce efficiently or rapidly and typically lack robust mechanical properties. These attributes tend to make them impractical for light-based additive manufacturing (3D printing) techniques such as stereolithography (SLA). Here, linear semi-crystalline, thiol-ene photopolymer systems will be discussed that form in seconds at low irradiation intensities and exhibit remarkable mechanical properties such as high elongations to break and toughness while also maintaining the capacity to melt and form a relatively low viscosity liquid for 3D printing applications as casting materials.

626272 Comings and Goings of Research and Education in Flow at Small Reynolds Numbers

Abstract:

In this brief talk, I'll summarize a few research themes from my early days (the "comings") of research on flows at small Reynolds numbers, and then describe how my group has returned to these themes (the "goings") in more recent years. The themes include sedimentation and aggregation of small particles, elastohydrodynamic collisions of wet particles, and drop interactions and deformation. In recognition of the Warren K. Lewis Award in Chemical Engineering Education from AIChE, I will also say a few words about the privilege of contributing to the education of students in the course of this research.

629031 Development and Incorporation of Kit-Based Experiments for Instruction of Fluid and Energy Transfer Laboratory Using Online and Hybrid Teaching Platforms

Abstract:

Traditional approaches in Chemical Engineering instructional activities focus on the scale-up of small experimental procedures starting at the Chemistry Lab scale, transitioning to pilot-scale experiments in Unit Operations laboratories for the subsequent projection of industrial-scale manufacturing operations. While most of large manufacturing processes have been established by following the aforementioned scale-up approach, Chemical Engineers are nowadays working on a plethora of applications involving multiple scales. Micro and nanofluidic devices combined with temperature control is an example depicting the intrinsic nature of small-scale technologies, but some industrial operations must be designed in the seemingly non-conventional direction thus taking certain processes down to a smaller scale. Most of experiments conducted in Unit Operations laboratories depend on equipment often comprised of various subcomponents and even though partially or fully automated, they will require the physical presence of experimenters due to technical and safety considerations. This is a limitation in the event of interruption of face-to-face instruction such as that originated by the ongoing COVID-19 pandemic, making online instruction highly challenging for laboratory-based classes, especially because of the decreased -or total lack- of hands-on learning outcomes. Motivated by the need of finding alternatives to keep the hands-on component in remotely-assisted classes while facilitating the understanding of fundamental concepts in Chemical Engineering, experimental kits were conceptualized, designed, and created with the objective of having one kit per student that could satisfy the MCS criteria: modular, cost-effective, and safe. Kits were designed to be comprised of various components, having a 3D-printed main device along with external sensors wired to Arduino-based microprocessors for real-time monitoring and data collection. The 3D-printed devices were designed according to Engineering considerations to allow for specific flow regimes, flow rates, pressure/temperature ranges, and other constrains associated to the small scale of these designs. The 3D-printed devices included a fluidic bench, a packed bed column, coaxial and shell & tube heat exchangers, and cylindrical connectors acting as pressure taps. In these devices, effective lengths and diameters were in the order of centimeters and built-in pipes and conduits in the order of millimeters. Ancillary kit components included fluid components such as flexible tubing of various diameters, aquarium pumps, valves, connectors and reducers, along with electrical components for mini circuits such as wires, resistors, transistors, etc. Plastic labware was also included for liquid collection and volume measurement, sensors for differential pressure and temperature, among other elements. Boxes containing kit components required for experiments were given to students, and laboratory manuals were created containing theory, system configuration guidelines, standard operation procedures, and links to videos and other resources. The Fluid and Energy Transfer Operations Lab (Unit Operations 1) was completely revamped for this new platform hence four experimental modules were created: Fluid Flow, Flow Characteristic Curves, Packed Bed Column, and Heat Exchangers. Preliminary assessments from remotely assisted, synchronous experiments demonstrate the feasibility of these kits to keep the hands-on component of the class while enhancing the understanding of Fluid Mechanics and Heat Transfer phenomena. Additionally, kits have demonstrated a high versatility for multiple experimental designs, expanded statistical analysis, troubleshooting strategies, enhanced team dynamics, and the opportunity foster the student creativity by means of alternative experiments. Furthermore, the use of these kits has been extended for the instruction of Unit Operations 1 under hybrid platform with experiments conducted by students in a design studio-type classroom, and simultaneously other students performing the same experiments online. The evaluation of this hybrid platform is currently underway but based on the overall great response from students, the establishment of kit-based and pilot-scale combined modalities in Unit Operations classes will also be targeted as an alternative to intercalate multiscale analysis to further enhance learning outcomes for students in laboratory-based classes.

630164 Remote Lab Operations: Development of a Take-Home Lab Kit for the Exploration of Fluid Flow and Pump Characteristics

Abstract:

The shift in teaching modality due to COVID required flexibility and adaptability by instructors and students alike. While some courses easily transitioned, hands-on lab courses faced greater challenges. For many, the solution was to create recorded demonstration videos of lab experiments. While this response meets basic requirements for student exposure to concepts, it falls short in the development of psychomotor and sensory skills.

In a typical unit operations course, students are exposed to bench scale industrial processes such as heat exchange, pumping and fluid flow, reaction kinetics, distillation, and adsorption. Students learn how to operate equipment (e.g. pumps) and use instrumentation (e.g. thermocouples and rotameters). Development of psychomotor and sensory skills require physical interaction with objects and stimuli from the environment. To accomplish this in the remote modality, a take-home laboratory kit was proposed. The challenge was to balance cost, durability, safety, and relevance to engineering phenomena. After prototyping a few models, the team selected a kit that demonstrated fluid flow characteristics using a submersible pump.

The kit was comprised of common, readily available items such as: an aquarium pump, clear vinyl tubing, a gallon plastic jug, and a flexible tape measure. It also included a power meter and ground fault circuit interrupter. The kit was stored in a clear plastic bin which doubled as the water reservoir for the experiment. From this simple kit, students could observe fluid flow characteristics (physical representation of the Bernoulli equation) and calculate head losses and power requirements. Ultimately, the students would model pump curves from their data and predict pump performance under varying physical conditions.

As course modality shifts away from remote, sustainability of this innovation was considered. The kit can continue to be used as a regular or make-up lab assignment. Additionally, the kit can be pivoted for use in sophomore and junior level classes to reinforce the concepts (like Bernoulli's equation) when they are first introduced to students. The difficulty level of the student activities can be adjusted to match their progression through the curriculum.

Here, we will present the details of acquiring parts, kit assembly, and sample student exercises and experiments.

626095 Combining Experimentation and Simulation in a Unit Operations Lab Course

Abstract:

In the chemical engineering curriculum, the Unit Operations and Process Design courses sit as capstones, where students combine theory with application. Typically, Unit Operations is a hands-on lab experience with pilot-scale equipment, while Process Design uses simulations to explore a larger, industrial-scale plant environment. In the second Unit Operations course at the University of Florida, students work on separations-based experiments, including continuous and batch distillations. These two columns share a control room and analytical equipment and, due to the Covid-19 pandemic and resulting physical distancing requirements, it was necessary to move one of the experiments online. Due to the conceptual difficulty of a transient batch process, the continuous distillation experiment was chosen. This presented an opportunity to enrich an existing lab experiment with some elements of the Design course. Students were asked to use old data and equipment specifications to simulate the continuous distillation column in Aspen HYSYS. Instructional videos were created by the Design instructor to guide students through data-based simulation, something not usually explored in the undergraduate curriculum. Once students had a working "base case" simulation, they experimented in a more open, "sandbox"-style way than when they are limited by the existing equipment. For example, they were able to study the effects of changing the number of trays in the virtual distillation column, something that is impossible with the physical equipment.

622747 Gaming for Engineering: Virtual Unit Operation Laboratory Via Unity

Abstract:

As COVID-19 crippled the world, the necessity of online teaching resources became more evident. Among different courses, laboratories and studios are still facing a greater challenge in preserving learning outcome of the course. The mainstream approach for the online lab-based courses are in-home kits, demonstration videos, and live-feed of admin-controlled labs. However, these solutions are still lacking few key ingredients. Replicating the hands-on experience, learning by trial and error, and on-site peer interaction which are bread and butter of the learning process in higher education. There is a growing interest in implementing simulated labs in the engineering curriculum. Simulated labs feature independence in experimental design, low cost experimental runs, interactive platform, and refreshing aesthetics. In particular, simulation is suitable for engineering labs that require high temperature/pressure and vacuum condition which triggers the cost associated with each run and upkeep of the instrumentation. As the cost of high-tech experimentations drop by adopting a similar concept in virtual reality world, conventional educational labs are remaining the bottle neck of the transition process towards fully online degrees. To create a replica of the physical lab in order to respond to current demand for distance learning and future demand for fully online programs, a Unity powered virtual lab has been developed to enable individuals, both instructors and students, to perform experimentations via an interactive platform, collect data, and apply analysis. In addition, conventional in-person labs are lacking the flexibility of the experimental design due to limited scenarios that are applicable and safety issues which could potentially prevent the promotion of critical thinking. Here, we have developed 4 modules of (1) Heat exchangers, (2) Pipes and Fittings, (3) Thermal Chamber, and (4) Pumps and Compressors as a replica of physical lab. The simulation has been tested by the same students that were present in the physical lab as well as video-called experimental sessions and their satisfaction has been measured compared. This tool can be utilized by schools for offering hybrid or fully online laboratory courses, individuals who are interested in chemical engineering world, and online teaching platforms that are considering online degree programs.

Talking

632380 From Nature to the Classroom: Culturing an Environmental Isolate in a Small-Scale Bioreactor for a Unit Operations Laboratory Course

Abstract:

Bioreactors are widely utilized in chemical industries, from large unstirred tanks in which yeast make ethanol to disposable-bag type reactors used to produce monoclonal antibodies. As such, bioproduction targets molecules for the pharmaceutical, food, cosmetics, and commodity chemical markets. Additionally, biological production offers a sustainable route of synthesis, often taking plant sugars that result from photosynthesis and converting them into chemicals. For the senior-level Unit Operations laboratory presented here, students characterized a two-liter bioreactor, which is commonly used to evaluate cell growth performance and production at a small-scale before moving to larger reactors. The bacterium studied is an environmental-isolate of *B. megaterium* (called SR7) that was isolated six-years ago from an underground CO_2 reservoir during my postdoctoral studies. Due to the industrial and societal relevance of the project, the novelty of working with a bacterium that few have studied, and the integration of topics from across the chemical engineering curriculum, students find the laboratory to be engaging and rewarding as they apply their previous knowledge to a challenging project. I will present the objectives for the laboratory, student-produced data, and ideas to implement this laboratory in other undergraduate programs.

623327 Unit Operations Experiments to Integrate Technologies and Experimental Design in the Curriculum: Mixing Applications

Abstract:

Mixing is part of many chemical engineering operations. Ineffective mixing can be costly and negatively impact product quality and safety. A general understanding of mixing operations can better prepare students to be strong contributors as they enter the profession. Combining a solid introduction to a technology, such as mixing, with experimental design serves to increase the breadth and depth of students' engineering education and makes students more marketable. For this work, mixing experiments in tank-type systems have been developed to familiarize students with key aspects of the technology. Experimental programs were developed using Six-Sigma methodologies for experimental design and data analysis. Students learned important mixing regime designations, impeller selection, and efficient gas introduction in liquid and liquid/solid systems including measurements of relative power demand. Students also developed proficiency with experimental design and experimental design and data analysis software. The experiments developed for this work were tested in a unit operations course. They had a beneficial influence on student learning and yielded positive student responses. These experiments are highly flexible and can be used with a wide range of students. They can be easily adapted for courses such as fluid mechanics, mixing electives, modules to teach experimental design to first and second year students, and K-12 outreach efforts.

631289 Strategies for Hands-on Activities in Undergraduate Engineering Curriculum in the Time of COVID-19 Pandemic

Abstract:

With the COVID-19 induced abrupt transition to online learning in undergraduate curriculum at West Virginia University in Spring 2020, new adaptations and delivery of core classes and hands-on laboratory courses in biomedical engineering had to be envisioned. This presentation will highlight both the challenges and the success of a pandemic-induced remote delivery of what are typically hands-on laboratory concepts and principles, Moreover, the presentation will emphasize the assessment of this transition on both educational outcomes and student successes. Specific cases will be related to three core courses, namely "Introduction to Biomedical Engineering", "Biomedical Engineering Laboratories" and "Human Physiology Laboratory". Discussions on the impact of synchronous lectures and recorded video demonstrations, inclusion of office hours and discussions boards in real-time or chat feature implementation will provide concrete examples of both students' integration as well as student feedback. Analysis of student performance with concepts evaluation as well as challenges associated with both the delivery and time management will also be included to help highlight how effective in terms of student learning outcomes, engagement and immersion the online delivery was. Based on our outcomes and on the summarized analysis of student surveys, we demonstrated that effective techniques could be developed and adapted to enhance student experience and create a more engaging and effective learning environment even in a virtual format. Further, we demonstrate how hands on activities integrated in the curriculum not only leads to better learning but further, how the challenges and future prospects for hands-on activities implementation could be used as a pipeline for dissemination of engineering related concepts and techniques to high school students to thus provide concrete examples of real-world integration and applications while allowing exposure to STEM disciplines.

632573 Effects of Temporal Parameters of Pulsed Electric Field Operation on Desalination Performance and Water Dissociation in Electrodialysis

Abstract:

Climate change, as well as the increased agricultural, municipal, and industrial consumption are exacerbating the ongoing pressure on limited freshwater supplies, leading to a global water scarcity crisis. Development of energy-efficient desalination processes for treatment of saline water is quintessential in reaching a sustainable water supply in the future. Among the existing desalination techniques, electrochemical membrane processes provide the opportunity for tunable and selective water treatment which allows for the adjustment of the treated water quality in accordance with the properties of interest for each specific application. Such a *fit-for-purpose* treatment approach can significantly improve the economics of desalination and its feasibility. Electrodialysis (ED) is an electrochemical membrane desalination process which utilizes an electric field to drive ions through cation and anion exchange membranes (AEMs and CEMs), alternately placed between two electrodes. The final products of ED are *diluate* and *concentrate* streams with, respectively, lower and higher salinity than the feed water. In an ED process, the higher transport number of counterions inside IEMs relative to the bulk solution leads to formation of concentration gradients in the adjacent boundary layers (BLs). Ionic concentration in the BL of the diluate channels decreases, while increasing in the BL of concentrate compartments. This phenomenon is known as "concentration polarization" which can intensify water dissociation in diluate channels and leads to formation of over-saturated regions in the concentrate channels. Water dissociation raises the hydroxide salt precipitation while over-saturated solutions indicate higher propensity for precipitation of sparingly soluble carbonate and sulfate salts. To improve the desalination performance, increase water recovery, and enhance the membrane lifespan, it is essential to mitigate organic and inorganic precipitation which lead to membrane fouling and scaling.

Pulsed electric field operation (PEF) is a low-cost approach for mitigation of membrane fouling and scaling through suppressing the concentration polarizations in the channels. In PEF-ED, the pulsing periods (t_{on}) with imposed electric field are followed by pausing lapses (t_{off}) with zero input current. The relaxation time provided during the pausing periods allows for the concentration in the boundary layers to return back to their bulk values, removing the concentration polarization formed during the pulse lapses. The performance of PEF-ED is controlled by adjusting pulsing frequency, duty cycle, and potential. Effectiveness of PEF operation for scale mitigation purposes relies on appropriate selection of pulsing potential and temporal parameters. Changes in pulsing conditions can affect the desalination rate, energy consumption of the process, and the pH variations in the system. In the literature, controversial results have been reported with respect to the pH variations in PEF-ED. While a number of studies have reported lower or identical pH changes in PEF-ED compared to the conventional ED (CED), several other investigations have determined intensified pH variations with PEF-ED. Minimizing water dissociation and pH changes are especially important for avoiding hydroxide salt precipitation and membrane scaling that are formed in basic pH. To reduce water splitting, ED is traditionally operated below the limiting potential at which the concentration of ions at the membrane-solution interface approaches zero. Existing literature does not address the effects of pulsing parameters on limiting potential which serves as the upper bound for the pulsing potential. Developing a systematic framework for selection of pulsing parameters that could improve the desalination rate while minimizing pH changes is an essential step, enabling the application of this approach for various water compositions.

We develop a 1-D transient model for PEF-ED by utilizing the fundamental equations applied to electrochemical systems including the Nernst-Planck, Navier-Stokes, and electroneutrality assumption. We further design and construct a bench scale PEF-ED set-up which facilitates the measurement and monitoring of voltage, current, conductivity, pH, and pressure drops in the system. Our theoretical and experimental analyses determine that limiting potential is greater in PEF-ED compared to that in CED for the same feed water salinity. Greater limiting potential achieved in PEF-ED allows for imposing higher pulsing potential which further improves the desalination rate. The higher limiting voltage results in minimal pH changes for a wider range of the imposed potentials, further reducing the possibility of hydroxide salt precipitation. The limiting conditions, however, depend on temporal parameters of pulsing and can vary at various t_{on}/t_{off} ratios. Our results illustrate the effects of PEF operation on pH changes and clarify the controversy in the literature. Our theoretical model and experimental procedure provide a systematic approach for tuning the pulsing parameters to improve the desalination performance in PEF-ED and mitigate water dissociation to avoid hydroxide salts formations.

629716 Catalyst Synthesis for Fuel Cell Application and Fundamental Understanding of CO_2 Reduction Mechanism

Abstract:

The electrochemical reduction of CO_2 is a promising way to produce fuels and chemicals. However, CO_2 electroreduction currently suffers from poor selectivity and stability and low energetic efficiency at practical reaction rates. To overcome some of the drawbacks and achieve a viable route for CO_2 reduction, the reaction should be split into a two-step process. Initially, CO_2 is reduced to CO, followed by a further reduction to the desired multi-carbons (e.g., C_2H_4, C_2H_5OH, etc.). Decoupling the process into two steps allows for CO electrolysis in highly alkaline conditions, enabling enhanced efficiency. Prior works have shown progress towards understanding the impact of electrolyte ions, pH, mass transport, temperature, and pressure on electrode activity and selectivity. A gas diffusion electrode (GDE) has been incorporated into a flow cell reactor to facilitate better transport and distribution of reactants and circumvent mass transport limitations. The development of an operando technique compatible with a flow cell reactor is highly desirable.

A promising technique that can provide insight reaction information in real-time is the flow electrolyzer mass spectrometry (FEMS). This analytical technique collects volatile products of the reaction in the immediate vicinity of the electrode. Using FEMS, we investigated the electrochemical carbon monoxide reduction reaction (eCORR) on polycrystalline copper and elucidated the acetaldehyde formation's oxygen incorporation mechanism. Combining FEMS and isotopic labeling, we showed that the oxygen in the as-formed acetaldehyde intermediate originates from the reactant CO, while ethanol and n-propanol contained mainly solvent oxygen.

As a part of my graduate studies, I have had the opportunity to conduct research on complex electrochemical processes. My doctoral dissertation focused on catalyst synthesis and electrochemical and physicochemical characterization. We identified the effect of catalyst support on the metal particle distribution and size, which has a broader impact on catalyst activity during electrochemical processes. Carbon black (CB), such as Vulcan XC-72, has been widely investigated and proposed as support for PEMFCs catalysts. Vulcan XC-72 exhibits an ideal high surface area and low production cost. However, organo-sulfuric impurities and the inaccessible metal particle sites inside the micropores may often lead to low catalytic activity values. These drawbacks turned the interest into alternative carbon-based supports such as carbon nanotubes (CNT), graphene (Gr), and Biochar (BC) which are all characterized by high specific surface area and high electronic conductivity. Comparing various carbon supports, we found that carbon structure and metal particle size play an important role in fuel cell electrocatalytic activity. Notably, carbon structures of high crystallinity (graphene and carbon nanotubes) and low pzc (point of zero charge) values facilitate the formation of smaller particles. These catalysts exhibit better performances in a PEM fuel cell under pure hydrogen and CO poisoning feeding conditions. On the other hand, Biochar and Carbon black exhibit amorphous phases and high values of pzc, resulting in catalysts with large metal particle size and low electrocatalytic activity.

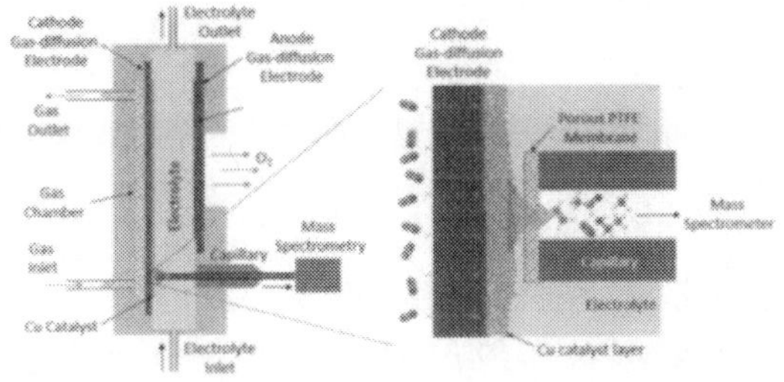
Electrolyte Outlet
Cathode Gas-diffusion Electrode
Anode Gas-diffusion Electrode
Gas Outlet
Gas Chamber
Gas Inlet
Electrolyte
Capillary
Mass Spectrometry
Cu Catalyst
Electrolyte Inlet
Porous PTFE Membrane
Mass Spectrometer
Capillary
Electrolyte
Cu catalyst layer

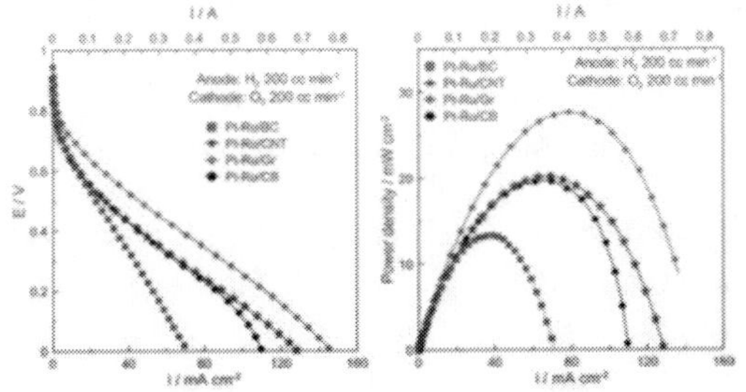
I / A
E / V
i / mA cm⁻²
Power density / mW cm⁻²

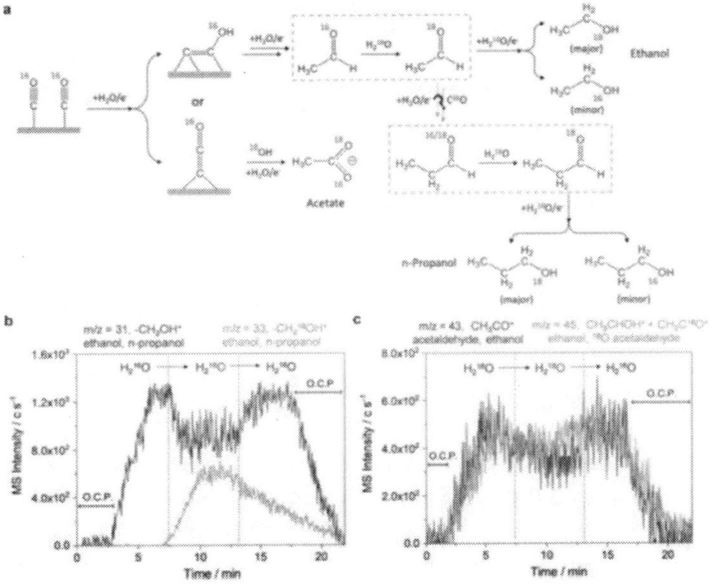
a
Ethanol
Acetate
n-Propanol
(major)
(minor)
b
c
O.C.P.
Time / min
MS Intensity / c s⁻¹

630148 Comparison of Thermal- and Electro-Catalytic Conversion of Biomass-Derived Oxygenates

Abstract:

Introduction

Hydrogenation and hydrogenolysis reaction studies have been carried out traditionally under high temperature and pressure in the presence of hydrogen gas fed into the reactor from an external source. Recently, electrochemical hydrogenation and hydrogenolysis has attracted a lot of attention since reactions can be carried in mild conditions (room temperature and ambient pressure) using water as the hydrogen source. Electrocatalysis can be combined with renewable energy resources (e.g. biomass and solar cells) to yield a sustainable path to producing valuable chemicals. An additional advantage of electrocatalysis is the added degree of freedom (i.e. electrode potential) that can be controlled to tune reaction pathways [1].

Despite the key advantages of electrocatalysis in the upgrading of biomass-derived oxygenates, it remains to be explored the differences in selectivities between thermal- and electro-catalytic hydrogenation and hydrogenolysis of polar oxygenated compounds. Here we report selective conversion of biomass-derived small oxygenates (C5 and C6 sugars) in both thermal- and electro-catalysis. Specifically, by combining experimental and first-principles density functional theory (DFT) calculations, we show that the reaction selectivities are distinct under these two conditions. That is, hydrogenation of furfural to form furfuryl alcohol is the major product in an aqueous-phase thermal reaction while 2-methyfuran is the primary and major product in electrocatalysis under more negative potentials. We further show that this difference in selectivities results from varied fractional charges transferred at the transition states along two reaction pathways.

Materials and Methods

Electrochemical reactions were carried out with an Agilent B2902A Precision Source/Measure Unit (SMU). Data was collected vs. Ag/AgCl reference. Copper foil (thickness 0.05 mm, Carolina) was used as the working electrode. Prior to each experiment, the surface was mechanically polished until no discoloration was visible then chemically polished using 0.1 M HCl solution to remove any oxides. Platinum foil was used as the counter electrode. The proton exchange membrane separating the two chambers of the elechtrochemical cell was immersed in 0.5 M H_2SO_4 for 24 hours prior to experiment. The liquid products from the anode and cathode chambers were extracted using Diethylether with a 1:1 ratio. 5 mL of pure acetonitrile was used to trap volatile products. The liquid product was filtered and analyzed by gas chromatography.

The plane-wave density functional theory calculations were carried out using Vienna Ab initio simulation package (VASP). The Perdew-Burke-Ernzerhof generalized gradient approximation exchange-correlation potential (PBE-GGA) was used. The model system is a supercell built based on four (4x4) repeated slabs of Cu and Ni (111). Explicit water molecules were added to model the Cu/water interface. The method is similar to our previous study of the Pd/water interface [2]. Reaction energies of all elementary steps were calculated using the model of computational hydrogen electrode. Electrochemical barriers at different electrode potentials were calculated based on fractional charges at the transition states.

Results and Discussion

The evolution of product distribution shown in Figure 1a as a function of time suggests that this reaction is of the form A → B → C → D, with B representing the intermediate furfuryl alcohol which undergoes a sequential ring rearrangement reaction in the presence of water and hydrogen at higher temperatures to form 2-cyclopenten-1-one and cyclopentanone. Products evolution as a function of electrode potentials in Figure 1b shows that, when a (H^+, e^-) pair drives the reaction, hydrodeoxygenation of furfural to form 2-

methylfuran is the preferred route [3]. We find in DFT calculations that, in a neutral aqueous solution, the hydrogenation of the carbon in the aldehyde group followed by the oxygen is the path of minimum energy to produce furfuryl alcohol (Figure 1c). However, in electrocatalysis, the dehydration path of furfural through the double hydrogenation of the oxygen (proton-assisted) followed by the hydrogenation of the carbon to make 2-methylfuran is preferred. The electrode potential controls the free energy of the dehydration step. A bias of -0.55V can be applied to alter the reaction path from hydrogenation (furfuryl alcohol route) to dehydration (2-methylfuran route).

Figure1 a) Product distribution in an aqueous-phase thermal catalytic reaction of furfural on 10% Cu/SiO_2 (P_{H2} = 300 psi), b) Product distribution in a liquid-phase electro-catalytic reaction on a Cu foil (0.5 M H_2SO_4), c) DFT-calculated free energies in liquid water, d) Comparison of reaction pathways of furfural in thermal- and electro-catalytic reactions.

References

1. Z. W.Seh, J.Kibsgaard, C. F. Dickens, I. Chorkendorff, J. K. Nørskov, T. F. Jaramillo. Science **355**, eaad4998 (2017).
2. Zhao Z, Bababrik R, Wang B, Resasco DE. Nature Catalysis. 2019Jan;2(5):431–6.
3. Bababrik, R., Santhanaraj, D., Resasco, D.E. J Appl Electrochem 51, 19–26 (2021).

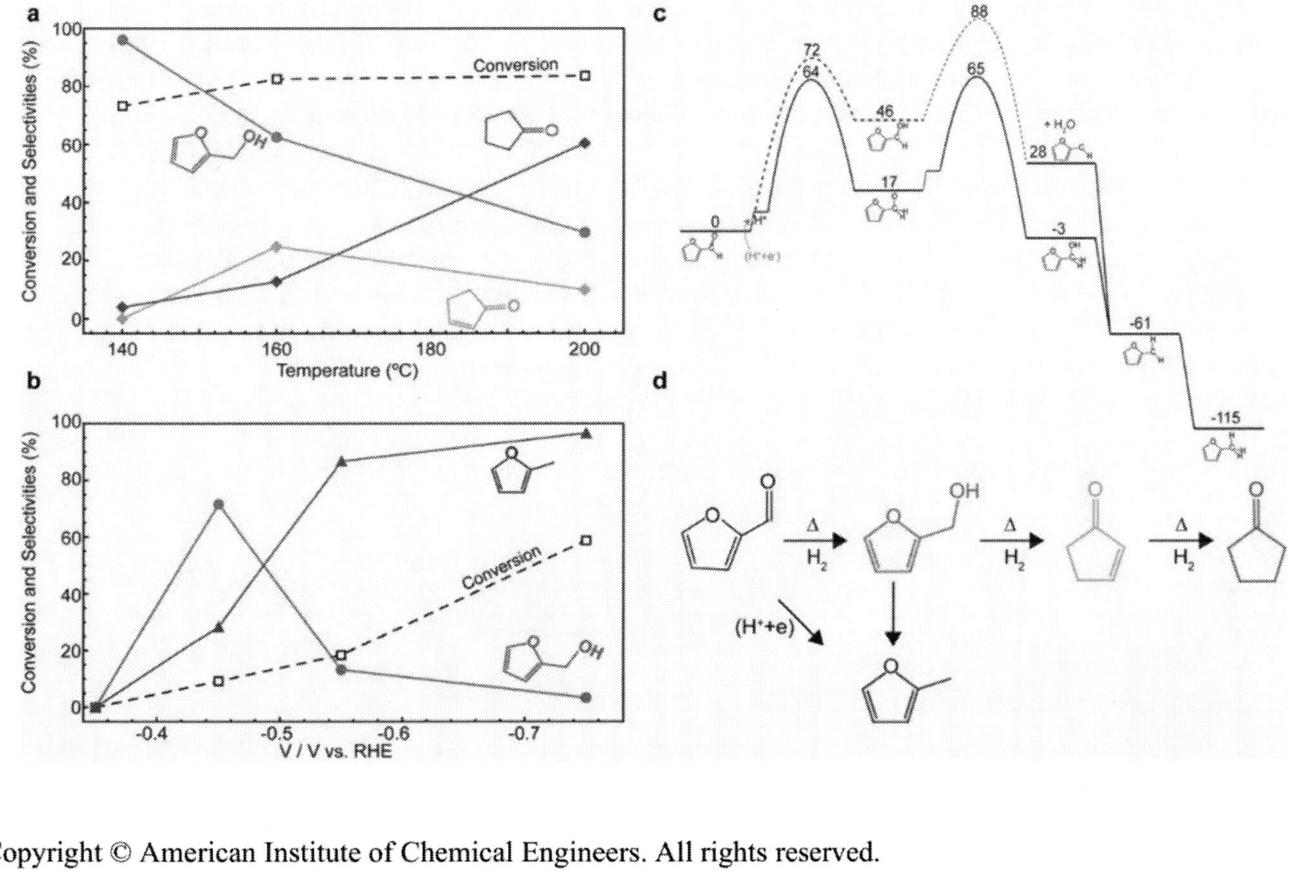

630421 Understanding the Extent of Ionic Dissociation and Ionic Conductivity in Model Thin Film Polymer Electrolytes As a Function of Different Side Chain Configurations

Abstract:

Numerous electrochemical systems depend upon ion-conducting polymer electrolyte materials used as membrane separators and electrode binders. In many instances, the rate of ion transport within these materials governs ohmic resistances and energy efficiency. A key challenge in ion-conducting polymer electrolyte materials for fuel cells, electrolysis, and electrochemical separations is robust and stable performance under a wide range of hydration values. Central to the rationale design of ion-conducting polymers is a fundamental understanding as to how chemical composition and macromolecular architecture influence thermodynamic and transport behavior of ions and water. In this presentation, I will present our work on studying the ionic activity coefficients and ionic conductivity of model thin film block copolymer electrolytes prepared via directed self-assembly and characterized via advanced metrology and molecular dynamics simulations. The preparation of model thin films commences with perpendicular alignment of self-assembled poly(styrene-*block*-2-vinyl pyridine) (PS*b*P2VP) lamellae to the substrate surface. The conversion of the non-ionic block copolymer into a block copolymer electrolyte occurred by alkylating the nitrogen in the P2VP domain with various halogenated reagents *via* a Menshutkin reaction. Alkoxy crosslinks and alkyl side-chain to pyridinium were prepared and compared against existing data sets on n-methyl pyridinium. New processing schemes were devised to introduce alkoxy and alkyl side chains post self-assembly without disrupting the nanostructure morphology. Atomic force microscopy (AFM) and X-ray photoelectron spectroscopy (XPS) confirmed retention of the structure after successful alkylation. The differences in activity coefficients with the various side chain configurations were determined via ion sorption experiments and swelling-deswelling behavior using a quartz crystal microbalance. Ionic conductivity was probed at different relative humidity percentages on interdigitated electrode substrates. Finally, molecular dynamics (MD) simulations were also performed to assess ionic dissociation and ionic conductivity on the BCEs with the various hydrophilic, alkoxy side chains and hydrophobic, alkyl side chains.

622853 Relationship between Aerobic Oxidation Catalysis and Electrochemical O_2 Reduction on Heterogeneous M–N–C Catalysts

Abstract:

Heterogeneous catalysts mediate the reduction of O_2 (ORR) and the oxidation of organic molecules both separately in electrocatalytic redox steps on-electrode, and in concert in thermocatalytic aerobic oxidations off-electrode. These disparate reactivity modes have been proposed to share key mechanistic features [1,2], but the generality of this concept remains unclear. Our group recently reported a fuel cell wherein electrochemically generated hydroquinone (H_2Q) molecules mediate thermocatalytic ORR on a heterogeneous metal–nitrogen–carbon catalyst (M–N–C) off-electrode [3]. Here, we interrogate the kinetics of H_2Q-mediated ORR on M–N–C to compare its mechanism with that of electrocatalytic ORR.

The ORR TOF (303 K, per total M) mediated by a tetra-sulfonated thioether-substituted H_2Q (E = 0.6 V vs. RHE) on a representative Co–N–C catalyst in the aqueous phase was quantified as a function of $[O_2]$, $[H^+]$, $[H_2Q]$, and [Q]. These kinetic dependencies ($[O_2]^1$, $[H^+]^0$, $[H_2Q]$ saturation, [Q] inhibition), together with an H/D KIE of 2.4 ± 0.4 in D_2O solution, are consistent with a cooperative mechanism (Panel (a)) that involves inner-sphere O_2 reduction by H_2Q [4,5], which is physisorbed on the surface near saturation coverages during catalysis, as corroborated by equilibrium adsorption isotherms. The inner-sphere mechanism of H_2Q-mediated ORR leads to a lack of correlation between on- and off-electrode ORR TOFs on a suite of M–N–C (Panel (b)), where Co–N–C catalyzes off-electrode ORR with higher TOFs than Fe–N–C, but not on-electrode ORR. ORR TOFs mediated by a series of H_2Q with varying redox potentials establish a linear free energy relationship (Panel (c)) whose slope deviates from the on-electrode ORR Tafel slope, further supporting the kinetic relevance of inner-sphere H_2Q-mediated O_2 reduction steps off-electrode. These insights provide guidance to design mediators and catalysts for ORR and aerobic oxidations.

References

[1] Wieckowski and Neurock, *Advances in Physical Chemistry* **2011**, *2011*, 1–18.

[2] Ryu, Surendranath, et al. *ChemRxiv* **2021**. 10.26434/chemrxiv.13830920.v1

[3] Preger, Stahl, et al. *Joule* **2018**, *2* (12), 2722–2731.

[4] Anson, Hammes-Schiffer, Stahl, et al. *J. Am. Chem. Soc.* **2016,** *138* (12), 4186–4193.

[5] Adams, Neurock, Flaherty, et al. *Science* **2021**, *371* (6529), 626–632.

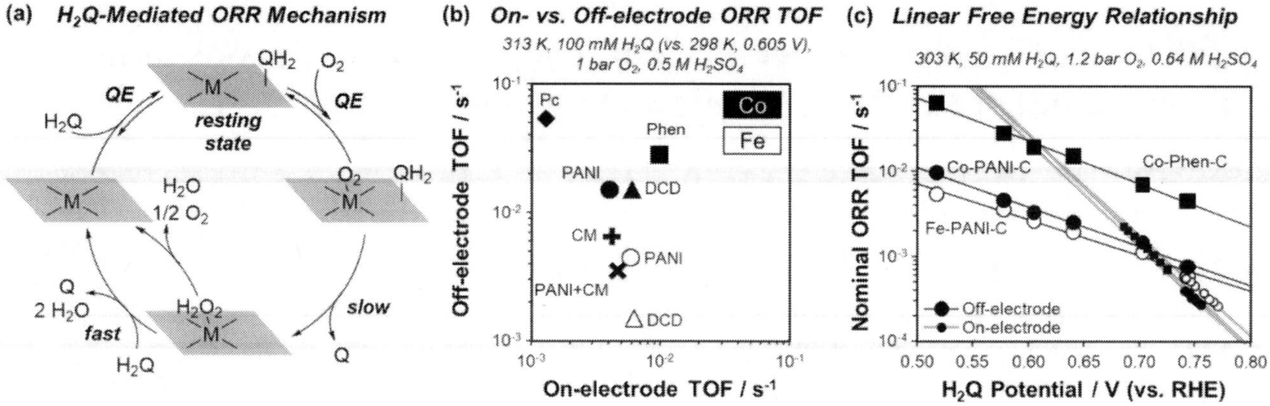
(a) H2Q-Mediated ORR Mechanism
resting state
QE
QE
slow
fast
(b) On- vs. Off-electrode ORR TOF
313 K, 100 mM H2Q (vs. 298 K, 0.605 V), 1 bar O2, 0.5 M H2SO4
Off-electrode TOF / s-1
On-electrode TOF / s-1
Co
Fe
Pc
Phen
PANI
DCD
CM
PANI+CM
(c) Linear Free Energy Relationship
303 K, 50 mM H2Q, 1.2 bar O2, 0.64 M H2SO4
Nominal ORR TOF / s-1
H2Q Potential / V (vs. RHE)
Co-PANI-C
Co-Phen-C
Fe-PANI-C
Off-electrode
On-electrode

623035 Electrochemical Reactions: Fundamentals Mapped from the Atomic to the Macroscopic Scale

Abstract:

In this presentation I will exemplify recent progresses in the theoretical description of electrified interfaces based on my current research activities. This will be discussed in the perspective of electrocatalysis and corrosion, with focus on the fundamental challenges in modeling and how these challenges can be addressed. In particular, I will present methods developed by me and collaborators for describing electrochemical barriers under different conditions – a long standing difficulty in computational chemistry. I will also showcase novel methods for linking atomic scale properties to the macroscopic behavior of electrochemical systems.

The motivation to my work stems from the grand challenges in our society, including the development of sustainable energy solutions for the future. Electrochemical methods are expected to be instrumental in this transition offering unique possibilities to control selectivity and improve efficiency in chemical processes. In order to materialize on this, it is essential to create nanostructured catalysts with tailored properties for a given application. In this endeavor, the development of novel theoretical methods that can capture the chemical behavior of complex nanostructures with atomic resolution will be essential for the screening of new catalysts candidates and in guiding towards rational design at the nanoscale.

To highlight the impact of our methods, I will present results for a number of applications of broad general interest. I will for instance discuss hydrogen generation through electrochemical water splitting – a central process for the emerging hydrogen economy, but also an ideal test case for methods in describing electrochemical barriers. In addition, I will demonstrate the use of our methods for obtaining detailed understanding of the mechanism of carbon dioxide reduction into valuable fuels and commodity chemicals over ideal and roughened copper surfaces. Finally, the issue of material stability will be addressed, in particular in the context of corrosion. As copper is a material of multifaceted use in the energy sector – such as catalysts in carbon dioxide reduction, as main material for the power grid, and as proposed material for storage of spent fuel from nuclear plants – it serves as a natural starting point. Here we leverage on the recent progress in atomic scale modeling, and link information at the atomic level to macroscopic observables through a series of linear relationships.

Putting my work on a timescale and in relation to the efforts of other groups around the world, it is obvious that we can only reach the ambitious goals of our society if we collaborate. I believe the work that I and my coworkers have carried out builds a foundation which will aid in accelerating the progress in the important field of electrochemistry, paving the way towards increased understanding of the fundamentals in electrochemical systems and catalysis.

632791 Integrating Student-Created STEM Comics into Chemical Engineering Education

Abstract:

In order to meet the demands of a growing student population, with an increasing number of sections and instructors for every core chemical engineering course, new curricula tools have been developed to help supplement current instruction and provide a stronger and more consistent educational experience for all undergraduates at Northeastern University. Of particular concern has been the need to improve student understanding of more complex, theoretical course concepts that do not have an obvious real-world application or are not traditionally represented in visual form.

Comics represent a growing effort in broader STEM education to utilize a visual-component to improve student interest, confidence, and understanding in the subject matter, and have been used in both formal and informal learning environments. While our efforts on using comics in chemical engineering education over the last five years have observed significant improvement in student engagement with the concepts and moderate improvement with their understanding, dedicated projects for several semesters have shown student-created comics to have a significant impact as well. Student performance on exams has dramatically increased through the presentation of comics as additional learning tools, and further, students have indicated significant improvement in their confidence in addressing the topics. This poster will discuss current efforts in the research and further assessment.

631883 Student-Generated Wikis As an Assessment Tool

Abstract:

The transition to remote learning in 2020 and 2021 required educators to change their approach towards both teaching and assessment. For the latter, traditional methods such as quizzes and exams had to be thoughtfully administered because students were no longer physically present with the instructor and had a wide range of resources that they could potentially use to aid their responses. To address this, student-generated wikis were assigned in several courses (i.e. Material and Energy Balances, Thermo I and II, and Computer Simulation) and used as an assessment tool during remote learning. Briefly, students were asked to distill down what they learned into the most important points and concepts, and could demonstrate this by incorporating content from the internet (e.g. webpages, multimedia, etc) in combination with their own personal explanations. This approach leverages a student's access to resources, relatively long time scale to complete the assignment, and generates a body of content that they may look back on as they progress through the chemical engineering curriculum. Student reaction to this assignment will be presented, as well as the activity's role with respect to traditional assessments, and continued implementation in traditional face-to-face courses.

632200 Gamification of Quantitative Problems to Increase Student Engagement

Abstract:

A web-based game, QRGames, involving quantitative problems that is cooperative within groups and competitive between groups has been developed. Students are exposed to conceptual, quantitative and judgement decisions in an environment which provides an engaging and relevant context.

In one form of the game, the context is a chemical production facility and teams are trying to maximize the health of this facility. The first phase involves conceptual or scaffolding questions in a traditional online gaming platform such as Kahoot, Gimkit or Quizziz. In the second phase, student receive additional points by solving multipart, quantitative, computer graded problems where each student receives a different version of the same problem. Student are encouraged to help teammates and are penalized if too much discrepancy in progress on the problem develops within a team. The progress of the teams and team members are displayed on a scoreboard.

In the final phase, the team decides the best way to spend the combined points to maximize the health of their production facility. For example, should the points be spent on an emergency shutdown system, upgrade the aging steam boilers, or fund research and development? Each purchasing decision can change the financial, environmental, and societal health of the facility. How these factors are weighted to sum to the team's final score is determined by the political environment that is finally and somewhat randomly selected.

Educational benefits include conceptual and scaffolding questions and peer to peer instruction on questions that are relevant to the game context. Additionally, student experience business type decisions with limited resources and tradeoffs between the financial, environmental, and societal health of an organization within an uncertain environment. Individual participation and contribution as well as team cooperation are incentivized. The game can be used as an effective review, an extracurricular activity, or a break from the normal routine of a class.

635068 In Honor of the 2020 Education Division Award Winner for Innovation in Chemical Engineering Education

Abstract:

For novel contributions to chemical engineering laboratory, design, engineering ethics, and first-year engineering instruction using game-based and narrative approaches.

635069 In Honor of the 2019 Education Division Award Winner for Innovation in Chemical Engineering Education.

Abstract:

Humor Applied to Enhancing Engineering Education, Communication, and Innovation.

Most of us believe that learning should be fun, and studies indicate that humor in the classroom can improve student engagement and satisfaction. However, humor can become a distraction in technical knowledge domains and worsen educational outcomes. Using a cognitive load theory framework, we have explored how a relatable proxy example can improve educational outcomes in technical areas such as science and engineering. If the example is humorous it can also increase student engagement, but the juxtaposition of technical analysis and relatable examples often provides the incongruity that produces humor with no effort. Humor in the form of humorous improvisation, was also utilized as the divergent thinking component in technical innovation exercises. We utilize humorous improvisation as a simulation in idea space to discover innovative ideas. This is analogous to using stochastic simulations to explore design and variable space in science and engineering.

636938 In Honor of the 2020 Education Division Award for Service to Chemical Engineering Education

Abstract:

For providing extensive service to engineering professional societies, delivering numerous educational workshops, and being a champion for diversity and inclusion in engineering education.

621410 Teaching Process Data Analytics and Machine Learning

Abstract:

Teaching chemical engineers to be effective in data analytics is as important today than in any time in the history of the discipline, and advances in sensor technologies, machine learning, and associated software have enabled applications that were not formerly possible. This abstract describes experiences with teaching process data analytics and machine learning over the last few years, including in (1) a new chemical engineering elective course in which the students come from chemical engineering, mechanical engineering, and engineering management, (2) a professional education course cotaught by chemical and mechanical engineering faculty whose learners primarily come with chemistry and chemical engineering backgrounds and many years of industrial experience, and (3) an undergraduate chemical engineering concentration in process data analytics with courses coming from a mixture of multiple disciplines including chemical engineering, computer science, economics, management. The overview will describe the challenges in teaching data science to chemical engineers and scientists and engineers working in the chemical and biotechnology industries, and strategies for overcoming those challenges.

622120 Teaching Big Data Science to Undergraduate Students in the University of Iowa

Abstract:

In this presentation, we will provide an overview of our course development and our experiences in teaching big data science and analytical skills to the undergraduate students in chemical engineering in the University of Iowa. These courses include: (a) Machine Learning (ML) and Artificial Intelligence (AI) course; (b) statistical analysis course; and (c) thermodynamic course. They are designed to be offered to students from freshman to junior years. We found that in general, the chemical engineering students are not getting used to use programming to analyze the data, in part because many courses they are taking focus more on the lab experiment while the analysis of the lab results is elementary and often done in Excel or by calculator. However, the students are willing to learn new analytical skills, and provided with well-designed teaching materials and problems, they can grasp the big data skills quickly. While the outcome of our experience has been encouraging, we also found there is clear lag in the field of chemical engineering to train our next generation with big data tools. This lag is part is due to the institutional tradition that emphasizes on the lab experiments and often deterministic way of thinking (with less consideration of probabilities), and in part due to the slowness in re-design the curriculum to reflect new development and ease access of ML and AI. We will provide examples to support the findings above, share the lessons we learned, and provide thoughts for next steps.

623878 Textile: Tutorials in Experimentalist Interactive Learning

Abstract:

Background:

Confocal microscopy is a common preclinical imaging modality to study a range of tissue features. Fluorescent image microscopy can provide insights into cell density, morphology, and cell-cell relationships. In our lab, we use immunofluorescent staining to investigate changes in cell and microenvironment features in response to development, injury, and treatment in the brain. We have developed image processing pipelines that threshold, segment, and quantify cell features. More recently, we have incorporated data science-enabled analysis and techniques into our imaging workflow. While the lab is comprised of researchers who are wet-lab experimentalists and data-driven analysts, only the second group has formal data science training. Wet-lab experimentalists can significantly benefit from having data science literacy to apply data science methodologies to their imaging datasets. However, there is no formal training regime to train wet-lab experimentalists in lab-specific data science methodologies. To address this training gap, we developed Tutorials in EXperimentalisT Interactive LEarning (TEXTILE) for data-science enabled fluorescent image processing.

TEXTILE is a module-based, semi-linear educational system developed with laboratory-based learning in mind. Modules combine data science fundamentals with specific biological imaging processing techniques. For example, we pair a data science module about data management with an image processing module on quantitative cell morphology characterization. The modular format allows us to train students on managing experimental data and pair that with quantifying features of cells that are relevant to their specific project of interest. Additionally, we designed TEXTILE such that the learning process is adaptable to starting skill level. Learning paths are based on entry knowledge but allow students to pick and choose their modules based on interest or previous experience. For example, someone with ample coding knowledge can skip the introductory Python modules and begin with the experimental image processing modules. Finally, we designed TEXTILE for a variety of degree levels. As a laboratory at a large public university, we engage with undergraduate and high school students, in addition to master's thesis and doctoral thesis students. We accomplished degree-level accessibility through the semi-linear system that assumes no prior knowledge in our research area. Since we already use TEXTILE to train high school students participating in laboratory research that have no previous knowledge of the subject topic, we aim to include more local high school students and expand TEXTILE for science education and outreach as well as in-laboratory education.

Methods:

TEXTILE is composed of topic-specific modules. Each module is designed around a data science topic, lasts an hour and a half, and contains four sections: the pre-module activity, main module, post-module reflection, and feedback form (Figure 1a). The instructor sends the pre-module activity to students for self-driven learning a week before teaching the main module. The pre-module's purpose is to prime students for the module's main lesson with a high-level activity. For example, in our data management module, students analyze a grocery shopping trip and decide where to store specific categories of food and perform a space evaluation on how much room in each storage location the food will take up.

The main module aims to teach the bulk material in digestible lectures broken up by interactive sessions. The main module's detailed structure consists of six parts: introduction presentation, interactive session one, main presentation workshop, interactive session two, concluding presentation, and a reflection session (Figure 1b). The introduction presentation introduces the module's topic, outlines the structure the main module will follow, and leads into the first interactive session. Interactive session one guides the student through an exploration of the module material. For coding-based modules, instructors lead interactive sessions with Jupyter Notebook. During interactive session one, the instructor walks students through the code's basics as students code along. After the students have written the code, the instructor begins the main presentation, which has two purposes: (1) to increase the students' connection between the pre-module activity and the module introduction and (2) to dive deeper into the technical aspects of the first interactive session. Once students understand the technical elements of the main module, they begin interactive session two, where they apply technical details of the lesson to further explore the code written in the earlier interactive session. After students have an opportunity to explore, the instructor gives the concluding presentation, which includes wrapping up the main takeaways, emphasizing the module's motivation, and thanking students for their participation.

Finally, the main module ends with a prompt for students to reflect on what they have learned and how their opinions about the module's topic have changed since the pre-module activity. The post-module reflection is completed individually by each student a day or two after the main module. The purpose is to provide a student-led review activity that encourages them to explore content from the main module on their own time to solidify concepts and build confidence. After the whole module is completed, students are sent a feedback form through Google Forms. With the feedback form, students reflect on their growth and the impact of each component of the module. The feedback is then implemented back into the module for the next cohort of students.

Results:

In the summer of 2020, the first cohort of students went through six TEXTILE modules of paired data science and image processing techniques used within the lab. The first cohort had 15 students: six high school students, four undergraduates, and five graduate students. The first cohort was taught live by a Ph.D. student from the lab's data-driven analysis team. The topics covered included GitHub and version control, introduction to data management, introduction to experimental data management specific to the lab, intro to image processing, thresholding, image segmentation, cellular morphology, and quantifying cell shape features. By the end of the first cohort, all students completed every TEXTILE module and the high school and undergraduate students developed individual projects using protocols learned over the summer. They each created Jupyter Notebooks for a self-led project with a brain cell morphology lit review, data management plan, and image processing pseudo-code. Many students from the summer 2020 cohort chose to continue researching with the lab and are currently developing these projects further.

In fall 2020, a small cohort of two undergraduate students explored the TEXTILE modules asynchronously. The students watched a recording of the main module and participated in asynchronous discussions with previous students and a Ph.D. student-instructor via Slack. We also provided these students with published image data that no one previously analyzed for cell morphology features. The students are currently completing retroactive analysis on this data using techniques learned from various TEXTILE modules.

For summer 2021, TEXTILE is expanded to include three modules on machine learning and two additional modules on experimental image acquisition and manual cell counting, incorporating student feedback from 2020. The main takeaways from student feedback included developing a formalized mentoring process for student projects and teaching via a hybrid live and asynchronous format.

Conclusions:

TEXTILE is a successful data science educational system for teaching data science techniques commonly used in experimental research labs. The TEXTILE system works for protocol training of fellow lab members, onboarding new students, and science education outreach. The combination of interactive sessions with instructor-led presentations gives students ample opportunity to work hands-on with real data from the lab in a guided environment. Students from the first cohort had the opportunity to develop their cell morphology research projects. In contrast, students from the second cohort took the lessons learned from TEXTILE and applied them for retroactive analysis of previously published data. The TEXTILE system is built to be robust and adaptable so that other laboratories can follow the methodology to develop training modules for their own data science techniques.

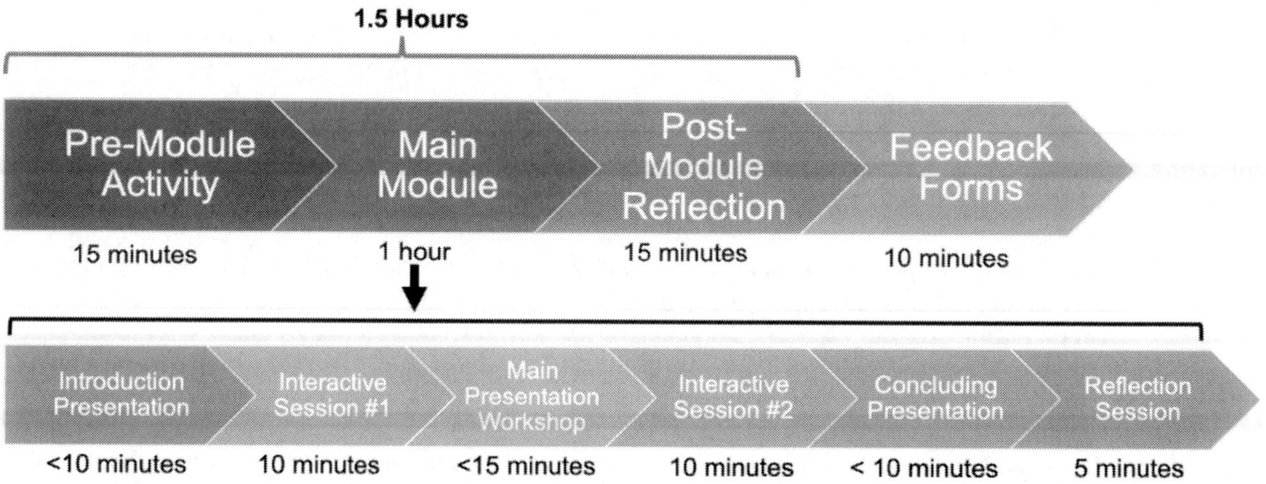

Figure 1. (a) Top row: TEXTILE process composed of four parts – pre-module activity, main modules, post-module reflection, and feedback forms. (b) Bottom Row: The six sections of the main module - introduction presentation, interactive session #1, main presentation workshop, interactive session #2, concluding presentation, reflection session.

622481 Teaching Artificial Intelligence to Chemical Engineers: Experiences from a 35-Year-Old Course

Abstract:

Given the recent success of artificial intelligence (AI) in a wide variety of applications, it has become quite clear that chemical engineering students need to become familiar with applied AI techniques [1]. There is great interest in the academic community in developing a course to address this important need. I have been teaching such a course to chemical engineering undergraduate and graduate students since 1986 [2], first at Columbia for three years, then at Purdue for about twenty years, and back again at Columbia now. It was mostly on *symbolic AI* in the early years, with an emphasis on *knowledge-based expert system* concepts and techniques, for various applications in chemical engineering such as design, diagnosis, control, optimization, and process safety. From the 1990s onwards, as machine learning techniques started showing promise, my course evolved to include topics such as clustering, neural networks, statistical classifiers, graph-based models, and genetic algorithms. The current version at Columbia has the first one-third of the course on *symbolic AI* and the rest on machine learning-based *numeric AI* concepts and techniques. It is mainly a project-based course where the students develop systems using AI techniques for chemical engineering applications. In the early years, the students programmed in AI languages such as LISP and OPS5, but now using the easier tools such as Python and TensorFlow. In this presentation, I will discuss my experiences in teaching this course, including the syllabus and the projects we have done over the decades.

Reference:

1. Venkatasubramanian, "The Promise of Artificial Intelligence in Chemical Engineering: Is it here, finally?", *Perspective article, AIChE Journal,* 65(2), pp. 466-478, 2019.

- Venkatasubramanian, V. "Artificial Intelligence in Process Engineering: Experiences from a Graduate Course," *Chemical Engineering Education*, Fall Graduate Education Issue, November 1986, pp. 188 – 192.

635310 Focus on What Matters – Advice on How to Spend More Time Doing the Things You Enjoy and Manage Your Academic Life

Abstract:

This talk shares my experiences, perspectives, mistakes and lessons learned in my journey through the academic ranks at The University of Washington. Modern faculty members are asked to wear many hats, and juggle an enormous number of responsibilities. At the same time, institutions often fail to provide clear guidance for new faculty on how they will be evaluated. Further complicating the situation is the reality that a faculty member's individual goals and priorities do not always have perfect alignment with those of their department and college. Through the lens of my own academic career, I will use this talk to provide a framework for holistic planning and evaluation of your progress in the professorate on timescales of 1, 3 and 5 years. The target audience is early career faculty, although trainees and recently promoted faculty may find some useful insights.

639147 Talk from Margo Vigeant

Abstract:

In this talk I'll concentrate on some immediate ideas for your teaching as well as some longer-term stuff for everyone to consider after you've successfully made it through a semester or two.

631475 Putting the Focus on Learning and Skill Development through Mastery Grading in Thermodynamics

Abstract:

Assessment, in particular the assigning of grades, is often one of the least rewarding ways in which instructors and students must interact with each other in the context of completing an academic course. Student expectations and the typical structures of assessment and grading both often lead to perceptions that grading can be arbitrary, does not match the student's own sense of achievement in the course, and that the instructor is somehow taking points or better grades away from students. In addition, the pandemic-induced switch to virtual and/or hybrid forms of assessment and examination have increased the prevalence and concern over cheating, through both greater accessibility using online resources and a heightened sense of need amongst a stressed student population.

In an effort to address these issues, the Mastery Grading assessment structure (also referred to as Specifications Grading) was adapted and implemented in two sections (68 total students) of a Chemical Engineering Thermodynamics 1 course in Fall 2020. The basic idea behind this methodology is to put the focus of the course on the development of skills and competencies, as opposed to the collection of points for a grade. As such, simple, timed assessments are built around identified learning targets (LTs) for the course and graded as either Satisfactory (S) or Unsatisfactory (U) (e.g. pass/fail). Students are then allowed to take these learning target assessments (LTAs) as many times as needed to demonstrate Mastery of that concept or skill based on a set of expected Specifications. Higher-level thinking and problem solving is then further assessed through Challenge Problems (CPs), which require the integration of multiple LTs to address a more complex problem. Students are allowed to spend unlimited and unrestricted time working through the CPs and may also re-submit them to correct errors following feedback and scoring. Grades are then earned as students progress through the semester in accordance with the level of demonstrated Mastery of course concepts and skills.

Upon completion of the semester, student survey responses indicated an overwhelming support for continued adaptation and implementation of this assessment structure. Students also self-reported a lower rate of cheating in this course compared to other courses that semester or in prior semesters. The average grades earned by students were comparable to grades assigned in prior iterations of this course, though a divergence was noted: while more students were able to put in extra effort to grow through initial failures (hence more higher grades), another subset of students found it more difficult to demonstrate certain levels of mastery (hence more lower grades as well). No longer was it possible for students to cram and muddle their way through 2-3 high-stakes examinations, often just happening to put together the right equations in the right order without understanding why. Instead, students were forced to continually engage with learning, practice, and assessment throughout the semester to grow their own capabilities and understanding. In summary, the Mastery Grading assessment structure was successfully adapted, implemented, and was well-received in a Thermodynamics 1 course.

Talbert, R., *Mastery Grading and Academic Honesty*, https://rtalbert.org/mastery-grading-and-academic-honesty/, July 20, 2020, Accessed 12 April 2021.

Nilson, L.B., *Specifications Grading: Restoring Rigor, Motivating Students, and Saving Faculty Time*, Sterling, VA: Stylus Publishing, 2014.

623031 Beyond Right and Wrong: Performance and Persistence in Solving Scaffolded, Randomized, and Auto-Graded Homework Problems

Abstract:

Positive life outcomes and learning are more strongly related to grit – combining passion and perseverance – than raw IQ. Measuring grit in chemical engineering education is addressed here. Auto-graded homework questions with randomized numbers and content provide frequent, formative assessments for students that created big data for instructors and researchers. Here, analytics are examined, and a model is constructed that combines student performance and persistence into a single score. This difficulty score shows potential to be used by faculty for creating assignments that are scaffolded, which has been shown to lead to greater learning. Over 800 questions and 170,000 attempts were studied from three cohorts of students (n=284) using a fully interactive online textbook, Material and Energy Balances zyBook. Problem sets consisted of multiple, scaffolded parts where easier levels preceded harder levels. Problem types included multiple choice as well as single and multiple numerical response. When students were allowed an unlimited number of attempts before a fixed due date, a median correct of 94% was measured. A difficulty score was established using both performance and persistence measures. Performance included both raw fraction correct and modified fraction correct, which accounted for only the students who attempted each level. Persistence was measured via attempts before correct. Based on a three-tiered point system, auto-graded questions were designated easy, medium, or hard, and cross correlation by question type and question order begins to validate the model.

624772 Data-Driven Continuous Improvement Strategies for Asynchronous Online Courses

Abstract:

The COVID-19 pandemic has affected ChE instruction in many ways, including the move of many courses to an asynchronous online format. Asynchronous online courses provide a unique highly-controlled arena for continuous improvement since course lectures, example solutions, etc. are pre-recorded; this means all students in a semester (and if no changes are made, in future semesters as well) receive identical course instruction and content. With this in mind a directed continuous improvement approach which assesses student performance on lecture-level learning objectives can be used to identify specific lectures or other course content which are weak or ineffective; such course content can then be critically reviewed and revised, and the impact of these revisions on student performance quantified. This talk will describe the application of this data-driven continuous improvement strategy for asynchronous online courses, including data collection, identification of weak course content, and results of application of the strategy.

629956 Effort-Based Grading of Homework

Abstract:

Problem-solving homework assignments are often graded on the basis of obtaining accurate answers to the questions posed in the problems. The resulting homework scores then count as some percentage of the students' course grade. This may motivate students to obtain the correct answers on their homework assignments, but in doing so, they may not learn how to solve those problems. Specifically, if a student finds a solution to the problem online or if a student finds a solution to a similar problem and mimmicks it, they will get the correct answer. However, when required to solve a similar problem without access to the internet and worked examples, the student is unable to obtain an answer. In this study, changing to effort-based homework grading was examined as means increasing the learning benefits of solving homework problems. The expectation was that by removing the pressure to obtain a correct answer, students would be able to focus on understanding how to get an answer. In three offerings of the course after changing the homework grading basis, neither the percentage of homework submitted, nor the effect of homework completion on course grade changed significantly. Suspecting that the students were still motivated by obtaining the correct answer, two additional changes were made. First, during in-class learning activities, the need to understand how to each type of problem and the differences between them was emphasized. Second, after submitting each assignment, the students were asked to reflect upon the process they used to solve the problem and write a brief statement on how well they understood the process and the reason for any mistakes they made while executing it. Adding these two changes on top of the effort-based grading did not change the percentage of homework submitted appreciably. However, the effect of homework completion upon course score increased by approximately 50%. That is, the slope of the trendline for course score as a function of homework completed increased from 0.29 to 0.44. The results suggest that students are motivated to obtain a correct answer to homework problems, and simply removing the reward for doing so is not sufficient to shift their motivation in the direction of deeper learning. However, the combination of explicitly telling them that their goals when solving homework should include being to develop understanding of how to solve different problem types, removing the penalty for not obtaining the correct answer and asking them to reflect upon their understanding after completing the homework resulted in a significant increase in the effectiveness of homework upon student learning.

624917 Authentic Assessment in Chemical Engineering Using Virtual Laboratories

Abstract:

This paper describes two case studies where technology-based virtual laboratories have been used to provide a tool for the authentic assessment of students. *Authentic assessment* uses real-world tasks that mirror the work required of professionals - in our case chemical engineers. This type of assessment requires students to engage in the type of reasoning and problem solving needed in the field. Thus, students must demonstrate deep understanding, higher-order thinking, and complex problem solving. Importantly, authentic interactions should also mirror the social interactions of engineering practice. Problems should be realistic, partly ill-defined (requiring the students to frame the problem), and complex with multiple solution paths possible. The work should be collaborative with students working on groups and other students' perspectives proving valuable resources.

I report on two instantiations of authentic engineering tasks mediated by a virtual reactor – one a fixed time two-hour exam and another an open-ended three- week project. I describe how these assessments align with the legitimate ways that knowledge is used in the profession. With this type of assessment, student teams can be evaluated on the performance of their engineering design, the accuracy that they report their findings, the quality of their writing, and the social interactions with their teammates – all facets of real engineering work. I will also briefly describe student reactions to this type of assessment and how it fit as instructional practice shifted to remote teaching during the COVID-19 pandemic.

621594 Over Half a Century of Protecting the Environment through Various Policies, Administrations, Successes, & Failures; An Overview of 50+ Years of EPA Regulations and Standards

Abstract:

Since the establishment of the U.S. Environmental Protection Agency (EPA) on December 2nd, 1970, the Agency has implemented a vast number of regulations to benefit the American people and the environment. Over the last 51 years, the EPA and its dedicated workers have endeavored to fulfil its mission to protect human health and the environment; the Agency has sustained its critical function through 51 years and 10 different administrations (6 Republican and 4 Democratic presidents). The change of administration leads to change and priorities, and at times new or rolled back regulations. One such case is the Bush-era fuel standard increasing the fuel standards (established in 1975) to a corporate average of 35 miles per gallon by 2020, further increased by the Obama-era fuel efficiency standard of 54.5 mpg by 2025, then replaced by the Trump-era standard requiring a small 1.5% increase in average fuel economy compared to the pre-Obama standards, and then upon the inauguration of the Biden-era, auto makers themselves began to support higher fuel efficiency standards as the Administration foreshadowed increasing the standards akin to the Obama-era. This presentation aims to provide an overview of current U.S. environmental regulations, where we currently are, where we have come from, and where we are likely to go in the future.

628185 Decarbonising Residential Heating: Cost-Benefit Analysis to Support Policy Design

Abstract:

The UK's legally binding targets to achieve "net zero" greenhouse gas emissions by the year 2050 may require full decarbonisation of residential heating. This will involve deploying low-carbon heating technologies (such as heat pumps, electric storage heaters, and/or various hydrogen heating technologies) across the country, including in the 24 million homes that currently use natural-gas-based central heating systems. A key challenge of this transition is that the capital and operating costs of low-carbon heating systems are often greater than current natural-gas-based systems. In addition, the UK has a wide variety of building stock, which defies a one-size-fits-all solution.

The aim of this study is to assess the suitability of various policy instruments to promote the uptake of low-carbon heating technologies. A cost-benefit model is being developed based on housing stock data from the English House Survey, performance data from field trials, and cost estimates from the literature. The model will be used to calculate the lifetime cost and levelised cost of heat for a wide range of households and the results analysed to consider how the costs of different low-carbon heating options vary with characteristics such as building type, number of occupants, thermal efficiency, and geographical location. In addition, an uncertainty analysis will used to evaluate the impact of different fuel prices, outdoor temperatures, and personal preferences and to identify the household types likely to be impacted most by these uncertainties.

The results of this study will be presented and discussed in relation to current policies designed to promote low-carbon heating. This study will consider what level of government support will be sufficient to promote low carbon heating over the current natural-gas-based heating systems and will identify if a targeted approach based on geographic location, building type, tenure type, or other characteristics would be more appropriate.

625673 Improving the Economics of Industrial Battery Storage: A Proactive Policy and Management Approach

Abstract:

Historically, implementation of battery energy storage has been proportionally higher in commercial and residential systems than industrial systems [1]. Economic and operational constraints of industrial facilities make battery installation close to infeasible. Industrial energy usage accounts for about 25% of total energy usage in the US. As the deployment of battery storage increases, lack of implementation in the industrial sector creates a gap, bridged by adapting control and policy strategies to constraints of industrial users. Projections indicate that battery prices will continue to fall, making it even more important to find ways to implement batteries in the industrial sector [2].

This research aims to show that the utilization of electric battery storage to decrease the energy costs and peak power usage of industrial energy users can be made attainable by identifying and closing the policy gap between affordability and utility of industrial battery installations. Similar studies have been done to improve economics and management of storage in residential installations [3]. Other studies have looked at regulatory design of energy storage coupled with renewable sources such as solar and wind [4]. Though these efforts have furthered the penetration of battery storage in the residential and commercial sectors, this same penetration has not been seen in the industrial sector from lack of analysis and specificity to the needs of industry. Policies specific to large electrical users in the industrial sector have also not been explored.

Load data from actual industrial facilities is used spanning a full year of typical manufacturing operations in which the daily electrical peak is about 2200 to 2500 kW. Two models are used, a battery model utilizing the facility load data to look at battery dynamics and effects of system load, and a financial model to analyze overall cost and effect of policy on project economics. The study identifies policies that are currently in place in parts of the US or similar policies that would allow batteries to be financially viable for industrial facilities.

[1] EIA, "Battery Storage in the United States : An Update on Market Trends," Washington, DC, 2020.

[2] W. Cole and A. W. Frazier, "Cost Projections for Utility-Scale Battery Storage," 2019.

[3] M. N. Sheha and K. M. Powell, "An economic and policy case for proactive home energy management systems with photovoltaics and batteries," *Electr. J.*, vol. 32, no. 1, pp. 6–12, 2019, doi: 10.1016/j.tej.2019.01.009.

[4] O. Zinaman, T. Bowen, and A. Aznar, "An overview of behind-the-meter solar-plus-storage regulatory design: Approaches and case studies to inform international applications (Report no. NREL/TP-7A40-75283)," *NREL Rep.*, no. March, pp. 1–76, 2020.

627260 How We Teach: Survey Results for Material and Energy Balances

Abstract:

Each year the Curriculum Survey Committee of the Education Division surveys institutions in the US and Canada over a portion of the undegradauate curriculum. The course for fall 2021 is Material and Energy Balances. Surveys will be collected in September, and results from the survey will be presented in this talk.

620718 MEB: The Sun in the Center of the ChE Universe

Abstract:

The material and energy balance (MEB) course is arguably the most important course in the chemical engineering curriculum with regard to both content and impact. It forms the foundation of the material that will be revisited, layer by layer, in successive ChE core courses. Students in MEB make the decision to continue in the major – or not – based on their experience in the course. Impressions formed in MEB frame the context of how students view chemical engineering as a discipline and a future profession. Best practices for teaching MEB as well as the course's implicit challenges will be discussed.

627265 Discussion: Issues and Best Practices in Teaching Material and Energy Balances

Abstract:

This time will be used for group discussion of issues and best practices in teaching material and energy balances.

633895 Overview of NSF Chemical, Bioengineering, Environmental, and Transport Systems Division (CBET)

Abstract:

The mission of the Chemical, Bioengineering, Environmental and Transport Systems (CBET) Division of NSF's Directorate for Engineering is to support innovative research and education in the fields of chemical engineering, biotechnology, bioengineering, and environmental engineering, and in areas that involve the transformation and/or transport of matter and energy by chemical, thermal, or mechanical means.

CBET research and education investments contribute significantly to the knowledge base and to the development of a 21st century workforce for major components of the U.S. economy, including chemicals, pharmaceuticals, medical devices, forest products, metals, petroleum, natural gas, food, textiles, energy utilities, alternative energy sources, microelectronics, and other sectors. Support for environmental research encompasses pollution prevention and remediation as well as life-cycle analysis.

This talk will highlight Foundation, Directorate, and Division level opportunities that may be of interest to the AIChE community of researchers.

633896 Chemical Process Systems Cluster Overview

Abstract:

The Chemical Process Systems cluster includes the: 1) Catalysis; 2) Electrochemical Systems; 3) Interfacial Engineering; and 4) Process Systems, Reaction Engineering, and Molecular Thermodynamics programs. Chemical processes are critical in the production of chemicals, materials, clean water, energy, pharmaceuticals, and other commodities. The Chemical Process Systems (CPS) Cluster's four programs support fundamental science and engineering research for the development of novel materials, mechanisms, and tools to improve the efficiency, resource utilization, and intensification of chemical processes. This cluster supports research seeking innovations in catalyst design, reaction engineering, electrochemical energy storage and chemical production systems, multiphase chemical separations processes, molecular thermodynamics, process control, optimization, and design, plasma processing systems, and sustainable energy conversion. The overarching goal of the CPS programs is to support basic research that improves the overall efficiency and product yields of chemical processes while reducing the size and complexity of process equipment and minimizing emissions. The CPS Cluster welcomes proposals that align closely with one of the four core programs, or that integrate multiple aspects of the core programs.

Catherine Walker, Chemical Process Systems Cluster

Carole Read, Electrochemical Systems

Robert McCabe, Catalysis

Christina Payne, Interfacial Engineering

Raymond Adomaitis, Process Systems, Reaction Engineering, and Molecular Thermodynamics

633897 Transport Phenomena Cluster Overview

Abstract:

The Transport Phenomena Cluster at CBET supports fundamental research on the transport of mass, momentum, energy, and species. There are four core programs in this cluster: Fluid Dynamics, Particulate and Multiphase Processes, Combustion and Fire systems, and Thermal Transport Processes. These core programs span across several disciplines including Chemical, Mechanical, Aerospace, Petroleum, and Environmental Engineering.

633900 Environmental Engineering and Sustainability Cluster Overview

Abstract:

Overview of the NSF CBET Environmental Engineering and Sustainability (EES) Cluster: Core Programs and Convergence Research Opportunities

The CBET EES Cluster is structured around three core programs: 1) Environmental Engineering, 2) Environmental Sustainability and 3) Nanoscale Interactions. This presentation will give an overview of the core programs of the CBET EES cluster. We will also highlight convergence research funded by the EES cluster with a focus on recent RAPID funded projects on wastewater-based epidemiology and the dynamics of pathogen transmission in the natural and built environment.

633899 Engineering Biology and Health Cluster Overview

Abstract:

The Engineering Biology and Health (EBH) Cluster consists of five programs that focus on the integration of engineering and life sciences to solve problems affecting humanity. These programs include Biophotonics, Biosensing, Cellular and Biochemical Engineering, Disability and Rehabilitation Engineering, and Engineering of Biomedical Systems. The EBH Cluster supports engineering research to improve human health and the understanding of biological and physiological systems. This includes efforts to promote novel and reproducible systems for detection and screening of biological analytes for health and environmental monitoring, modeling in cells and tissues, stimulating advancements in synthetic biology and protein engineering toward improved biomanufacturing, and enhancing quality of life of persons with disabilities through fundamental engineering advances. The scope of the EBH Cluster encompasses multiple topics that cross programs both within CBET and throughout NSF.

633901 Interactive Question and Answer Session with NSF Program Directors

Abstract:

Interactive Question and Answer Session with NSF Program Directors

The mission of the Chemical, Bioengineering, Environmental and Transport Systems (CBET) Division of NSF's Directorate for Engineering is to support innovative research and education in the fields of chemical engineering, biotechnology, bioengineering, and environmental engineering, and in areas that involve the transformation and/or transport of matter and energy by chemical, thermal, or mechanical means.

CBET research and education investments contribute significantly to the knowledge base and to the development of a 21st century workforce for major components of the U.S. economy, including chemicals, pharmaceuticals, medical devices, forest products, metals, petroleum, natural gas, food, textiles, energy utilities, alternative energy sources, microelectronics, and other sectors. Support for environmental research encompasses pollution prevention and remediation as well as life-cycle analysis.

This talk will highlight Foundation, Directorate, and Division level opportunities that may be of interest to the AIChE community of researchers.

639806 Academic Ableism and its Alternatives

Abstract:

About the speaker: Dr. Jay T. Dolmage is a professor of English language and literature at the University of Waterloo. He is committed to disability rights in his scholarship, service, and teaching. His work brings together rhetoric, writing, disability studies, and critical pedagogy. He is the author of *Academic Ableism: Disability and Higher Education* (2017) and *Disability Rhetoric* (2014). He is also the founding editor of the *Canadian Journal of Disability Studies*.

639807 Increasing Participation of Students with Disabilities in STEM

Abstract:

Students with disabilities are a significantly underrepresented demographic in the STEM disciplines. To help alleviate this inequality, we, at the University of Delaware, have developed the unique summer outreach program "Research Experience for Undergraduates Science and Engineering Leadership Initiative." This innovative, NSF-funded program encourages undergraduates with physical, psychological, or learning disabilities to take an active role in the scientific endeavors at a research-intensive University. This integrative measure is designed to empower the participants to pursue research-related careers and professions related to STEM disciplines. In our presentation, we will not only discuss the details of our program but aim to give experience-based, practical advice on how to best accommodate students with disabilities in research laboratories and ways to implement similar programs in other institutions. We will discuss the advocacy work of the American Chemical Society Committee on Chemists with Disabilities.

About the speaker: Dr. Sharon Rozovsky is a professor of biochemistry at the University of Delaware. She is one of the faculty directors of the Science & Engineering Leadership Initiative, a research experiences program designed for the advancement of students with disabilities who are interested in careers in chemistry, biochemistry, or aligned fields. She served on the Committee on Chemists with Disabilities of the American Chemical Society in 2018–2020.

633902 Proposal Writing Tutorial

Abstract:

Interactive tutorial on how to write a high-quality proposal to receive research funding from NSF. The session is suitable for new faculty member as well as senior faculty members.

633903 Interactive Breakout Panels

Abstract:

Interactive Breakout Panels for research funding from CBET.

The mission of the Chemical, Bioengineering, Environmental and Transport Systems (CBET) Division of NSF's Directorate for Engineering is to support innovative research and education in the fields of chemical engineering, biotechnology, bioengineering, and environmental engineering, and in areas that involve the transformation and/or transport of matter and energy by chemical, thermal, or mechanical means.CBET research and education investments contribute significantly to the knowledge base and to the development of a 21st century workforce for major components of the U.S. economy, including chemicals, pharmaceuticals, medical devices, forest products, metals, petroleum, natural gas, food, textiles, energy utilities, alternative energy sources, microelectronics, and other sectors. Support for environmental research encompasses pollution prevention and remediation as well as life-cycle analysis.This talk will highlight Foundation, Directorate, and Division level opportunities that may be of interest to the AIChE community of researchers.

631711 Integrating Anti-Racism and Social Justice throughout the Chemical Engineering Curriculum to Create More Conscious Engineers

Abstract:

Traditional chemical engineering homework and in-class problems are often theoretical or simplified and lack societal context. However, a design made without context may only work for the demographics of the person(s) creating the design and unmarries the work of chemical engineers from the tangible impacts the field has on the world. In these contextless problems, students aren't given the opportunity to consider the positive or negative impact of their design on different demographic groups. Working towards de-centering western civilization, being sensitive to oppressions, integrating more cultural context and ethics into teaching, and reflecting on the role of identity in and outside chemical engineering will encourage critical thinking with an equity lens [1]. In light of the ongoing COVID-19 pandemic and anti-Black and anti-Asian racism and violence, institutes of higher learning, which have historically exacerbated systemic inequalities that disproportionately impact minorities and people of color, need to critically rethink and redesign curriculum to be more widely inclusive, accommodating, and thoughtful of social justice. Incorporating context, such as the context in human-centered design or service learning, provides relevancy for students and improves learning, engagement, and increased community engagement [2]-[3]. There are already examples of institutions and faculty incorporating social justice and environmental considerations into their curriculum through human-centered design and service learning and a few who are incorporating context into class examples [4]. However, we have yet to see evidence of incorporating anti-racist and social justice principles in a systematic and curriculum-wide way within a chemical engineering context. Thus, we present an approach to normalize anti-racism and social justice ideals throughout curriculum, rather than presenting them as one-off, disconnected ideas, such that these become virtues students gain and take with them into their professional careers after graduating. Our work is part of a larger effort of the Anti-Racism, Diversity, Equity, and Inclusion (ARDEI) Committee in the Chemical and Biological Engineering Department at Northwestern University, which is comprised of undergraduate students, graduate students, and faculty. The larger goal of the committee is to work together to build and contribute to a Department that is diverse, equitable, and inclusive through actionable ideas that the Department can pursue together to make lasting change in its culture. Collectively we tackled the initiative of incorporating anti-racism and social justice into the undergraduate curriculum. We aimed to find a low-cost, low-barrier-to-entry method for faculty to include anti-racism and social justice into the classroom. Specifically, we advocate for the inclusion of these principles into homework problems and course project designs at all course levels. In order to achieve this goal and get faculty support, we decided to create a bank of example problems, spanning freshmen to senior courses, incorporating critical thinking or example problems with the context of anti-racism and social justice. We subsequently shared these example problems with the faculty and invited them to bring example problems from their courses to a workshop where we help them incorporate these principles into the desired problem. We also aimed to incorporate anti-racism and social justice principles into lectures. For example, in the senior design course, topics of environmental and social justice are worked into lectures along with the term-long process design project, where traditionally economic and technical evaluations were emphasized, now environmental and societal impact evaluations are required for decision making in the feasibility study. Though most faculty are not trained in anti-racism work, we hope to alleviate this challenge and lower the barrier by working with them to create these examples. Additionally, we must carefully manage our work to ensure the anti-racism and social justice contexts and materials included do not lead to unintentional harm. We hope our work will serve as an example that other institutions can follow to incorporate anti-racism and social justice into the chemical engineering curriculum. However, we importantly note that this should not be the only actions taken by an institution to participate in anti-racism and social justice; otherwise the inclusion of anti-racism into courses will be seen as hypocritical if the institution is not also working towards making more equitable, inclusive, safe learning spaces. This work can be seen as one of many important steps in creating these improved learning spaces and ideally lead to more anti-racist and socially just engineers.

[1] D. Riley, "Engineering and Social Justice," *Synthesis Lectures on Engineers, Technology, and Society*, vol. 3, no. 1, pp. 1–152, Jan. 2008, doi: 10.2200/S00117ED1V01Y200805ETS007.

[2] Y. Wang and R. Rodgers, "Impact of Service-Learning and Social Justice Education on College Students' Cognitive Development," *NASPA Journal*, vol. 43, no. 2, pp. 316–337, Jul. 2006, doi: 10.2202/1949-6605.1642.

[3] C. Baillie, A. Pawley, and D. Riley, "Developing human-centered design practices and perspectives through service-learning / Monica E. Cardella, Carla B. Zoltowski, and William C. Oakes," in *Engineering and Social Justice: In the University and Beyond*, West Lafayette, IN: Purdue University Press, 2008.

[4] L. J. Hirshfield and H. B. Mayes, "Incorporating Inclusivity and Ethical Awareness into Chemical Reaction Engineering," *Chemical Engineering Education*, vol. 53, no. 4, pp. 215–219, 2019.

632796 Peer Mentorship Via Bridge (Building Relationships for Identity Development and Growth in Engineering) Program at University of Illinois at Chicago during Pandemic

Abstract:

Nearly 50% of students who begin as engineering majors do not graduate with an engineering degree. Retention is even lower for underrepresented minorities (URM), especially those who are first-generation and from low SES backgrounds. Thus, continual strategies should be employed to continue to retain students, especially in universities with high URM populations. Mentorship can be a valuable asset for students by providing the proper support and encouragement to accomplish their goals. Several studies have shown the benefits of peer mentorship on retention rates, GPA, and the development of one's self-belonging and engineering identity.

However, while previous studies have examined the impact of such programs on retention and engineering identity, the impact of these programs in a virtual learning atmosphere has not been as readily investigated. To examine this, a student-led peer mentorship program was instituted at the Department of Chemical Engineering at the University of Illinois at Chicago (UIC) at the start of the Fall 2020 year, coinciding with the start of a nearly complete transition to virtual learning.

To implement the program, a survey was sent out to upperclassmen students (in their final two years of undergraduate study) within the chemical engineering department at UIC. In addition to assessing interest, this survey also asked for pronouns, year in school, and professional/extracurricular interests of the potential mentors. Respondents to the survey were then asked to attend a brief training session in the summer to go over their responsibilities as a mentor. At this time, a survey was sent out to incoming freshmen assessing their pronouns and professional/extracurricular interests. Mentors and mentees were then matched by similarities in these two parameters.

Mentors were encouraged to reach out to their mentees after the matching had occurred, and were instructed to check-in and answer any questions that arose throughout the semester. After a semester of the program, a survey on engineering identity, belongingness in the engineering community, as well as involvement in engineering events were distributed to both mentors and mentees.

A second follow-up survey will be conducted at the end of the school year. A preliminary analysis of the data has shown that while no mentees felt like part of the chemical engineering community at the start of the program, 45% did by the end of one semester of the program. Mentors have expressed that their role has improved their communication skills greatly, and prepared them for being a better mentor later in their career.

Overall, this program has been a platform where both the mentors and mentees have benefitted. The hope is that this program can be implemented in other universities as well to allow more students to identify as engineers, thrive as engineering students, and prepare them for their future engineering discipline careers.

632937 Developing Engineering Identity Via Integrating up-to-Date Industry Problems into Mass and Energy Balances Course

Abstract:

Chemical engineering field has been expanding over the last few decades and now includes many diverse fields such as pharmaceuticals, renewable energy, biotechnology, foods, etc. Along with the evolution of the field, the competencies and skills needed by industry have increased. Now, practicing chemical engineers are expected to have expertise in many different fields and developed various technical and professional skills. While chemical engineering education continues to evolve and incorporate all these new aspects into the curriculum, many of the courses remain content-driven, inter and intra-personal skills are not developed, and faculty do not have access and resources for industry relevant problems. As a result, students are less motivated to pursue chemical engineering degree and to complete chemical engineering programs. Thus, the impact of industrial engagement in engineering courses and programs on professional identity formation, student engagement, learning and motivation, needs to be better understood.

The overall aim of this project is to develop chemical engineering students' professional identity prior to their graduation by integrating both academic and industrial workplace competencies within engineering programs. This pilot program in the Department of Chemical Engineering at the University of Illinois at Chicago (UIC) focuses on the sophomore-year and involves introducing industry-relevant up-to-date problems into one of the first courses in the chemical engineering sequence (*CHE 210- Material and Energy Balances*). During Spring 2021 semester, the project team collaborated with two professional engineers (industrial mentors) from different industry areas to integrate current practices into the target course (*CHE 210: Materials and Energy Balances*). The project team and industry mentors worked on designing up-to-date application-based problems/projects to be implemented in the homework assignments used by the course instructor. Problem/project was delivered to students in a video format recorded by the mentor, including introducing themselves, their company, and the project. Each problem/project had four stages to be completed in a month. Each part of the problem/project related to the concepts taught in the course and was systematically introduced to students over four weeks during Spring 2021. For example: one project was on carbon recycling including these steps: simple cost calculation, reactor conversion, separation, recycle and heat integration.

The project team investigated the effects of introducing video-recorded industry-relevant problems on students' professional identity formation and self-efficacy development. Consented students were surveyed prior to and following the intervention. A small sample (n=6) of randomly selected students were interviewed to capture more insightful data. Furthermore, course instructor, industry mentors and teaching assistants (TA) were interviewed on the impacts of the intervention. This presentation will describe the integration efforts and the analysis of data collected in Spring 2021.The Spring 2021 activities constitute a baseline condition for ongoing research efforts to increase industry engagement through multiple forms of contact with industry mentors. The project team plans to continue analyzing the impacts on students' professional identity and self-efficacy in Fall 2021 and Spring 2022 when industry mentors will interact more directly with the students enrolled in the course.

The knowledge and insights gained from the research will help engineering programs to analyze important aspects of industry engagements, the feasibility of implementation and impact on student outcomes and faculty professional learning. The results will allow engineering educators to improve the next generation's skillset and better prepare them for the workforce.

629658 Reverse Engineering Assessments: A Project-Based Assignment to Substitute for Exams

Abstract:

Instructors are often challenged to design exams to appropriately evaluate student knowledge and mastery of course material. In the time of online learning, administering exams and evaluating student knowledge has become even more challenging. Students at home have additional hurdles to overcome while taking exams including internet connectivity issues, maintaining a quiet environment during the exam, downloading and uploading work, and completing the exam in a similar amount of time as when classes were taught in-person. Furthermore, proctoring has become more difficult, especially for large classes where all students cannot be seen in one frame. In light of all these challenges, Reverse Engineering Assessments (REAs) were created as an alternative to exams. With an REA, students are assigned prompts on various course topics where they have to write and solve their own problems. Students are given a week to complete the REA and are given freedom in how to demonstrate their knowledge and mastery of material in class. Students have enjoyed the flexible scheduling in their assessment and the various options to prove knowledge and mastery. Student submissions demonstrated appropriate levels of knowledge comparable to material covered on an exam.

629381 Developing Online Teaching Resources to Enhance Accessibility in Higher Education

Abstract:

Engineering courses include a lot of equations and abstract concepts. These equations are often coded in LaTeX and they are distributed as PDF copies to students. These PDF versions are not readily accessible to all students. Some students who use assistive technologies like screen readers rely on reading out loud or alternative representation of content. Using student-staff partnership, we have developed online teaching resources to improve accessibility as well as facilitate learning by providing an interactive and engaging online environment.

Based on student responses to a survey on relevance of supplementary resources, we found that course textbooks provided by instructor forms the primary source of information outside of lectures. Therefore, we transformed the traditional PDF based course notes for 2nd year Engineering Mathematics course into a website with LaTeX rendering and screen reader compatibility. We have added appropriate resources and links to allow for easy navigation between chapters and used uniform colour coding to separate theorems, definitions, and examples. We also included drop-down solutions for examples to provide students an opportunity to try the examples before looking at the solutions. The embedded images can be zoomed in and the text and equations are set up to fit the screen size and hence, can be read via devices including phones, tablets, and laptops. We included Google analytics functionality to assess student engagement with the website and used few surveys to assess student perception of the implemented changes. This transformation can allow easy integration of visual aids and several online resources to elucidate abstract mathematics concepts to students.

While course notes are essential resource for learning concepts, students rely on problem sheets and past exam papers for applying these concepts. Problem sheets can also serve as formative assessments if graded and feedback is provided in a timely manner. Often these are PDF based problem sheets and suffer from similar accessibility-related limitations. Additionally, owing to large class sizes, it can be difficult from an instructor's point of view to provide detailed feedback to students. To overcome these challenges, we implemented an online math-aware platform, WeBWork, to deliver problem sheet and exam-style questions. This platform can recognize mathematical notations and provides instant feedback on the correctness of the solutions. After a deadline, the detailed solutions are presented within the question as a hyperlink. Students reported increased number of attempts on a question and found this technology-based intervention to be beneficial for their learning. Data analytics from the platform also allowed instructors to identify difficult concepts and address them in real-time either during a lecture or by providing more support.

Both a course website and an online math-aware platform were implemented for the first time during the academic year 2020-21 during the pandemic-led restrictions and were well-received by the students with highly positive feedback and reviews. Future work will be aimed to incorporate student feedback and improve the quality of content and to embed more relevant resources to make Mathematics more interactive and accessible to all students.

629563 Development of an Emulsion Experiment for Process Engineering Lab for Undergraduate Education in Colloids

Abstract:

Emulsions are homogenous mixtures of two immiscible liquids where one phase is finely dispersed in the other. They are found in a variety of consumer productions including food, pharmaceuticals and cosmetics. Although, this is a phenomenon that chemical engineering students often encounter after graduation, students often have a weak background in this area due to the absence of a required emulsions or colloids course in the chemical engineering curricula. One opportunity to increase student exposure and retention in this field is the integration of an open-ended emulsion experiment in the senior process engineering laboratory. As part of the half-semester experimentation cycle students are divided into teams to design an emulsion experiment, test a variety of variables, and have weekly team oral pop quizzes.

629782 Deployment of Machine Learning Models in Pharmaceutical Development

Abstract:

The advantage of model driven decisions in research organizations has long been recognized and applied in the pharmaceutical research and development. From the conceptualization of the Quality by Design framework as a series of mathematical constructs [1,2] to the application of digital twins [3]. Machine Learning methodologies are widely applicable and efficient tools to construct mathematical models of relevant pharmaceutical processes enabling process optimization, robustness characterization, and failure rate estimation. However, the widespread adoption of these approaches requires a combination of streamlined availability, and institutional know-how to interpret and incorporate the results in the decision-making process.

This talk addresses different approaches to leverage emerging and stablished tools in Machine Learning to pharmaceutical development applications by comparing the adoption of specific algorithms and computational platforms. Specifically, a generic workflow for data exploration analysis, design of experiments, feature selection, and model exploration has been developed and deployed with a combination of cloud computing and Jupyter notebooks. In our experience this approach has been more successful at integrating a model-based decision making culture in process development than previously attempted alternatives.

[1] Peterson, J.J., 2008. A Bayesian approach to the ICH Q8 definition of design space. *Journal of biopharmaceutical statistics*, *18*(5), pp.959-975.

[2] Garcia-Munoz, S., Luciani, C.V., Vaidyaraman, S. and Seibert, K.D., 2015. Definition of design spaces using mechanistic models and geometric projections of probability maps. *Organic Process Research & Development*, 19(8), pp.1012-1023

[3] Chen, Y., Yang, O., Sampat, C., Bhalode, P., Ramachandran, R. and Ierapetritou, M., 2020. Digital Twins in Pharmaceutical and Biopharmaceutical Manufacturing: A Literature Review. *Processes*, *8*(9), p.1088.

626098 Unsupervised Learning from Sets of Data Using Contrastive Latent Variable Models

Abstract:

In unsupervised learning, dimensionality reduction is an important tool for data exploration and visualization. Because these aims are typically open-ended, it can be useful to frame the problem as looking for patterns that are enriched in one dataset relative to another. These pairs of datasets occur commonly, for instance a population of interest vs. control or signal vs. signal free recordings.However, there are few methods that work on sets of data as opposed to data points or sequences. Here, we present a probabilistic model for dimensionality reduction to discover signal that is enriched in the target dataset relative to the background dataset. The data in these sets do not need to be paired or grouped beyond set membership. By using a probabilistic model where some structure is shared amongst the two datasets and some is unique to the target dataset, we are able to recover interesting structure in the latent space of the target dataset. The method also has the advantages of a probabilistic model, namely that it allows for the incorporation of prior information, handles missing data, and can be generalized to different distributional assumptions. We describe several possible variations of the model and demonstrate the application of the technique to de-noising, feature selection, and subgroup discovery settings.

627704 Energy Dispersive X-Ray Hyperspectral Image Analysis and Chemometrics for Catalyst Characterization

Abstract:

Multi-dimensional images with chemical information embedded at pixel-level (so-called Hyperspectral Imaging) is the next generation of Image Analysis and Machine Vision. This advanced technology has attracted the attention of different industry sectors in the last two decades, ranging from metalurgic, mining and pharma, to high-value-added manufacturing industries such as biotechnology and biomedical imaging. Recently, the combination of Scanning Transmission Electron Microscopy (STEM) and Energy Dispersive X-ray spectroscopy (EDX) has been developed to characterize the surface area of catalysts in the chemical industry. The measurements are done by scanning the electron beam over the sample and at the same time measuring the X-ray spectrum – so-called spectral imaging [1]. The spectral dimension of the hyperspectral images consists of X-ray counts acquired at different energy channels. Chemical data is corrupted with noise, which comes from the time-dependent arrival of discrete particles on the sensor. This noise typically follows a Poisson distribution, which represents the probability of occurrence or events (X-ray counts) during a given period of time.

The application of Multivariate Image Analysis (MIA) techniques and Multivariate Curve Resolution (MCR) models becomes essential for the analysis of EDX hyperspectral images. This contribution proposes a modeling framework that permits segregating hyperspectral X-Ray images into simpler images (so-called Distribution Maps, DM's), which can be directly related to each of the chemical compounds present in the mixture [2]. From these DM's, chemical-textural score images (SI's) are further obtained. Finally, from all these DM's and SI's, different types of features, such as quantitative, morphological or textural can be extracted and combined into new data structures [3]. This new source of information is used to build multivariate statistical models for process understanding and prediction purposes at a MIA image-based level. This approach allows us to study similarities and differences between and within types of catalysts -i.e. different samples of the same catalyst across batches, and across locations, and the potential effects on quality properties of interest. To illustrate the chemometric framework for catalyst characterization, STEM-EDX images of real industrial catalyst will be used.

[1] R. Haswell, D. W. McComb and W. Smith, Preparation of site-specific cross-sections of heterogeneous catalysts prepared by focused ion beam milling, Journal of Microscopy (2003) Vol. 211, Pt 2, pp. 161-166

[2] J.M. Prats-Montalbán, A. de Juan, A. Ferrer, Multivariate image analysis: a review with application, Chemometrics and Intelligent Laboratory Systems, 107: 1-23, 2011.

[3] C Duchesne, JJ Liu, JF MacGregor, Multivariate image analysis in the process industries: A review, Chemometrics and Intelligent Laboratory Systems 117, 116-128, 2012.

634389 Applications of machine learning in production enhancement and field monitoring in Oil & Gas industry

Abstract:

Due to the fast advancement of machine learning technologies and availability of diverse types of data collection platforms, we have been seeing the growing machine learning applications in a wide spectrum of industries, spanning from emerging industries, e.g., autonomous driving, to traditional ones, such as Oil & Gas (O&G).

In this talk, we will discuss three different machine learning applications in the O&G industry. In specific, we will discuss a machine learning-based approach for understanding the multiphase flow system studied in the experimental fluid mechanics; which shows the potential of developing a complimentary methodology to assist the first principal physics; secondly, we will describe a deep learning-based object detection system that can be used to monitor the operation field or the natural environment, with the goal to improve operation safety and meet the regulation requirement ; Thirdly, we will introduce the content understanding of Process & Instrument Diagrams (P&IDs) using computer vision techniques.

Each of the above studies represents a unique usage scenario in the value chain of O&G industry. We would like to demonstrate the diverse types of machine learning applications and discuss their respective technical challenges that O&G industry presents.

AUTHOR INDEX

AIChE
120 Wall Street, FL 23
New York, NY 10005-4020

ISBN 978-1-7138-5700-6

9 781713 857006

CURRAN ASSOCIATES INC.
proceedings.com